The IMA Volumes
in Mathematics
and its Applications

Volume 55

Series Editors
Avner Friedman   Willard Miller, Jr.

Springer

New York
Berlin
Heidelberg
Barcelona
Budapest
Hong Kong
London
Milan
Paris
Tokyo

Institute for Mathematics and
its Applications
IMA

The **Institute for Mathematics and its Applications** was established by a grant from the National Science Foundation to the University of Minnesota in 1982. The IMA seeks to encourage the development and study of fresh mathematical concepts and questions of concern to the other sciences by bringing together mathematicians and scientists from diverse fields in an atmosphere that will stimulate discussion and collaboration.

The IMA Volumes are intended to involve the broader scientific community in this process.

Avner Friedman, Director
Willard Miller, Jr., Associate Director

* * * * * * * * * *

IMA ANNUAL PROGRAMS

1982–1983	Statistical and Continuum Approaches to Phase Transition
1983–1984	Mathematical Models for the Economics of Decentralized Resource Allocation
1984–1985	Continuum Physics and Partial Differential Equations
1985–1986	Stochastic Differential Equations and Their Applications
1986–1987	Scientific Computation
1987–1988	Applied Combinatorics
1988–1989	Nonlinear Waves
1989–1990	Dynamical Systems and Their Applications
1990–1991	Phase Transitions and Free Boundaries
1991–1992	Applied Linear Algebra
1992–1993	Control Theory and its Applications
1993–1994	Emerging Applications of Probability
1994–1995	Waves and Scattering

IMA SUMMER PROGRAMS

1987	Robotics
1988	Signal Processing
1989	Robustness, Diagnostics, Computing and Graphics in Statistics
1990	Radar and Sonar (June 18 - June 29)
	New Directions in Time Series Analysis (July 2 - July 27)
1991	Semiconductors
1992	Environmental Studies: Mathematical, Computational, and Statistical Analysis
1993	Modeling, Mesh Generation, and Adaptive Numerical Methods for Partial Differential Equations

* * * * * * * * * *

SPRINGER LECTURE NOTES FROM THE IMA:

The Mathematics and Physics of Disordered Media

Editors: Barry Hughes and Barry Ninham
(Lecture Notes in Math., Volume 1035, 1983)

Orienting Polymers

Editor: J.L. Ericksen
(Lecture Notes in Math., Volume 1063, 1984)

New Perspectives in Thermodynamics

Editor: James Serrin
(Springer-Verlag, 1986)

Models of Economic Dynamics

Editor: Hugo Sonnenschein
(Lecture Notes in Econ., Volume 264, 1986)

George R. Sell   Ciprian Foias
Roger Temam
Editors

Turbulence in Fluid Flows
A Dynamical
Systems Approach

With 17 Illustrations

Springer

George R. Sell
School of Mathematics
University of Minnesota
Minneapolis, MN 55455
USA

Ciprian Foias
Department of Mathematics
Indiana University
Bloomington, IN 47405-4301
USA

Roger Temam
Laboratoire d'Analyse Numérique
Université Paris-Sud
Bât. 425
91405 Orsay
France

Series Editors:
Avner Friedman
Willard Miller, Jr.
Institute for Mathematics and
   its Applications
University of Minnesota
Minneapolis, MN 55455
USA

Mathematics Subject Classifications (1991): 76-06, 58F12, 76F20, 58F39, 35B32, 60Gxx

Library of Congress Cataloging-in-Publication Data
Turbulence in fluid flows : a dynamical systems approach / George R.
   Sell, Ciprian Foias, Roger Temam, editors.
      p.   cm. — (The IMA volumes in mathematics and its
   applications ; v. 55)
      Includes bibliographical references.
      ISBN 0-387-94113-4 (acid-free)
      1. Turbulence—Congresses.   2. Navier-Stokes equations—
   Congresses.   I. Sell, George R., 1937-   .   II. Foias, Ciprian.
   III. Temam, Roger.   IV. Series.
   QA911.T843   1993
   532'.0527'01515352—dc20   93-5272

Printed on acid-free paper.

Production managed by Bill Imbornoni; manufacturing supervised by Jacqui Ashri.
Camera-ready copy prepared by the IMA.
Printed and bound by Edwards Brothers, Inc., Ann Arbor, MI.
Printed in the United States of America.

9 8 7 6 5 4 3 2

ISBN 0-387-94113-4 Springer-Verlag New York Berlin Heidelberg
ISBN 3-540-94113-4 Springer-Verlag Berlin Heidelberg New York   SPIN 10538013

The IMA Volumes
in Mathematics and its Applications

Current Volumes:

FOREWORD

This IMA Volume in Mathematics and its Applications

TURBULENCE IN FLUID FLOWS:
A DYNAMICAL SYSTEMS APPROACH

is based on the proceedings of a workshop which was an integral part of the 1989-90 IMA program on "Dynamical Systems and their Applications."

We thank Shui-Nee Chow, C. Foias, Martin Golubitsky, Richard McGehee, George R. Sell and R. Temam for organizing the meeting. We especially thank G.R. Sell, C. Foias, and R. Temam for editing the proceedings.

We also take this opportunity to thank the Office of Naval Research, the Army Research Office and the National Science Foundation whose financial support made the workshop possible.

Avner Friedman

Willard Miller, Jr.

PREFACE

The phenomenon of turbulence in fluid flows is a basic issue of science which has implications over a broad spectrum of applications in modern technology. In order to be able to better understand this matter, one needs to obtain a deeper insight into the dynamical properties of the solutions of the Navier-Stokes equations. These equations, which are the equations of motion of incompressible, viscous fluid flows, have been the object of many investigations over the last 60 years, since the pioneering work of J. Leray in the 1930s.

A common feature of the theory of turbulence arising in the dynamical systems approach is to identify the turbulence with the long-time dynamics of the solutions of the Navier-Stokes equations. One of the earliest conjectures about the onset of turbulence was offered by Hopf and Landau, who sought to describe the onset of turbulence as growing out of a series of higher-order bifurcations in the global attractor for these equations. While some features of the Hopf-Landau conjecture are not widely accepted these days, it is the case that the connections between turbulence and the dynamical complexities of the global attractor offer a challenging framework for trying to understand this important physical phenomenon.

It is known, at least in the case of the two-dimensional Navier-Stokes equations that there is a global attractor and that this attractor has finite dimension. Furthermore, simple examples can be constructed which demonstrate that, under suitable conditions, the dimension of the attractor can be arbitrarily large. This suggests that the full picture of the long-time dynamics of the Navier-Stokes may involve many different degrees of dynamical complexity.

The Institute of Mathematical Applications workshop on a Dynamical System Approach to Turbulence in Fluid Flows was one of a trio of workshops which closed the year-long program on Dynamical Systems and their Applications. The papers contained in this volume represent various approaches for studying the interrelated concepts of turbulence and long-time dynamics of the Navier-Stokes equations and related problems.

George R. Sell

Ciprian Foias

Roger Temam

CONTENTS

APPLICATION OF AN APPROXIMATE R-N-G THEORY, TO A MODEL FOR TURBULENT TRANSPORT, WITH EXACT RENORMALIZATION

MARCO AVELLANEDA* AND ANDREW J. MAJDA†

Abstract. The important practical problem of "eddy diffusivity" and renormalization for turbulent transport is discussed. Then a simple model for turbulent transport with rigorous renormalization developed recently by the authors is described. The simple form of the model problem is deceptive; the renormalization theory for this problem exhibits a remarkable range of different phenomena as parameters in the velocity statistics are varied. Thus, the model problem is an interesting test problem for renormalized perturbation theories. In this paper the approximate R-N-G method of Yakhot and Orszag is applied to this exactly solvable model problem and developed in a detailed fashion. The predictions of the approximate R-N-G theory are compared with the exact renormalization theory. In one of the four different regions of nontrivial renormalization the R-N-G theory is exact both for predicting the anomalous scaling exponents and the Green's function for the equation for eddy diffusivity. The reasons explaining this spectacular success are given in Section 4. For the other three regions, the anomalous scaling exponents and Green's function predicted by R-N-G are a rather poor approximation to the actual renormalization theory. However, in the one region adjacent to the mean field boundary, the R-N-G theory predicts the first order Taylor expansion of the scaling exponent in agreement with the expected behavior of an ϵ-expansion procedure from critical phenomena.

SECTION 1: INTRODUCTION

Fully developed turbulence involves velocity fields which have a continuous range of excited space and time scales which induce substantial additional dissipation in high Reynolds number fluid flows due to the development of energetic small scales (see [9]). How can one assess the effect of the motion of the arbitrarily many smaller length scales on the large length scales without resolving the motion of the small scales in detail? This question is an extremely important practical problem since even with great advances in storage and speed for the next generations of supercomputers, it will not be possible to resolve all of the continuous range of length scales in a practical application involving turbulent fluid flow ([15] [26]). The proposed answers to this question involve theories of "eddy diffusivity". Such theories range from those based on dimensional analysis such as Prandtl's classical mixing length theory to recent attempts utilizing extremely sophisticated renormalized perturbation theories (see [15], [26]) involving ideas borrowed from field theory and statistical physics. The recent book by McComb [19] contains a detailed description of these approaches.

*Department of Mathematics, Courant Institute, New York, University, New York, NY 10012. ·Partially Supported by Grants NSF DMS-9005799 ARO DAAL03-89-K-0039 AFOSR 90-0090

†Department of Mathematics and Program in Applied and Computational Mathematics, Princeton University, Princeton, NJ 08544. Partially Supported by Grants NSF DMS-9001805 ARO DAAL03-89-K-0013 ONR N00014-89-J-1044

Instead of discussing theories of eddy diffusivity for the Navier-Stokes equations, we develop these issues for the simpler problem of advection-diffusion of a passive scalar by an incompressible velocity field which is described by the equation

$$\frac{\partial T}{\partial t} + (v \cdot \nabla)T = \kappa_0 \Delta T$$

(1.1)

$$T|_{t=0} = T_0(x)$$

where the incompressible velocity field, $v(x,t)$ satisfied div $v = 0$ and $\kappa_0 > 0$ is the diffusion coefficient. We focus the discussion here on the case of (1.1) where the velocity field v involves a continuous range of excited space and/or time scales and admits a statistical description. Applications from fully developed turbulence for (1.1) include predicting temperature profiles or tracking pollutants in the atmosphere while the case when $\vec{v}(x)$ is a steady velocity field with many length scales occurs in the diffusion of tracers in heterogeneous porous media ([13]). The problem in (1.1) obviously has its own practical importance and is routinely used in the turbulence community as a simpler statistical model problem for the Navier-Stokes equations.

Here in the first part of this paper we present a careful mathematical formulation of the basic problems and goals of a theory for eddy diffusivity for the turbulent transport equation in (1.1). Then we present an instructive model problem for turbulent transport proposed and analyzed by the authors ([1], [3]) where the renormalization theory for eddy diffusivity can be found exactly. The simple form of this model problem is deceptive; the rigorous renormalization theory for this problem as described below exhibits a remarkable range of different phenomena as parameters in the velocity statistics are varied. This model problem provides a rigorous and unambiguous test problem for the wide variety of renormalization theories which typically utilize partial summation of divergent perturbation series ([16], [25], [10], [24], [14], [27]). We illustrate this facet of the rigorous exactly solvable model problem in this paper.

The recent R-N-G perturbation method of Yakhot and Orszag ([26], [27]) has attracted considerable interest. In the final parts of this paper, we develop the R-N-G method of Yakhot and Orszag in complete detail for the model problem including an exact solution of the ϵ-expansion procedure. We have attempted a pedagogical discussion of these methods on the model problem for readers without a background in R-N-G methods for critical phenomena. We conclude the paper with a detailed discussion of the predictions of this approximate R-N-G theory as compared with the exact answers for the model problem. Comparison of other approximate renormalized perturbation theories such as those involving weak coupling and Kraichnan's D.I.A. ([15]) as well as R-N-G on the exactly solvable model problem are presented elsewhere in joint work of the authors ([2]).

The remainder of this paper has the following:

SECTION 2: TURBULENT TRANSPORT AND THE EXACTLY SOLVABLE MODEL

2A): The Problems of Eddy Diffusivity and Infrared Divergence for Turbulent Transport. We begin by describing the problem of computing effective large scale transport equations for advection-diffusion by velocity fields with Kolmogoroff statistics. We regard the use of the exact velocity statistics implied by the Kolmogoroff hypothesis as a convenient concrete model for the velocity statistics to illustrate the subtle difficulties with turbulent transport; experiments to date and the Kolmogoroff theory agree reasonably well in the inertial range but it is unknown whether discrepancies in the agreement are due to the inadequacy of this theory or experimental inaccuracy.

With L_0, the integral length scale and $\bar{U}$, the typical velocity of the energy containing but non-universal fluid motions, the Reynolds number is given by $Re = \frac{\bar{U}L_0}{\nu}$ with ν the viscosity of the fluid; turbulent fluid flows with $Re \nearrow \infty$ are the primary interest here. The Kolmogoroff hypothesis in d-space dimensions (for $d = 3$) asserts (see [9]) that there is a well-defined dissipation length scale, L_d, so that as $Re \nearrow \infty$ the velocity spectrum has a universal form for wave numbers, k, in the range $L_0^{-1} < |k| < L_d^{-1}$ given by

$$(2.2) \qquad \left\langle |\hat{v}(k)|^2 \right\rangle = C_0 \bar{\epsilon}^{2/3} |k|^{1-d-5/3} .$$

Here $\hat{}$ denotes Fourier transform, $\langle \cdot \rangle$ denotes ensemble average over velocity statistics, $\bar{\epsilon}$ is the mean dissipation rate, and C_0 is a universal constant. The energy spectrum is assumed to vanish for $|k| > L_d^{-1}$ or decay very rapidly while the velocity is not universal on length scales larger than L_0. For the moment, we have assumed that the velocity field in (2.2) is steady for simplicity in exposition. Consider the advection-diffusion equation in (1.1) with the incompressible velocity field having the energy spectrum in (2.2). Assume that the initial data at time $t = 0, T_0(x)$, varies substantially only on length scales larger than L_0; the objective of theories of "eddy-diffusivity" modelling in this context is to determine an effective equation of motion involving variations of $T(x,t)$ on length scales larger than L_0 and for times at least comparable to the large scale eddy turnover time, $\bar{t} = L_0/\bar{U}$. We non-dimensionalize the equation in (1.1) by utilizing the dissipation length scale $L_d = (\nu^3/\bar{\epsilon})^{1/4}$ and the dissipation time scale $t_d = (v/\bar{\epsilon})^{1/2}$; using the relation $\bar{\epsilon} \cong \bar{U}^3/L_0$, one finds that

$$(2.3) \qquad L_d = (Re)^{-3/4} L_0 , \quad t_d = (Re)^{-1/2}\bar{t} .$$

With this non-dimensionalization and the identification $\delta = (Re)^{-3/4}$, the advection diffusion equation (1.1) assumes the form,

$$(2.4) \qquad \frac{\partial T}{\partial t} + v_\delta \cdot \nabla T = \kappa_0 \Delta T$$

$$T|_{t=0} = T_0(\delta x) , \quad \delta \ll 1 .$$

Here κ_0 is the Prandtl number; for most fluids κ_0 is a quantity comparable to unity. With (2.2), the rescaled velocity field $v_\delta(x)$ has the energy spectrum.

$$(2.5) \qquad \left\langle |\hat{v}_\delta(k)|^2 \right\rangle = \begin{cases} C_0|k|^{1-d-5/3} & , \quad \delta < |k| < 1 \\ 0 & , \quad \text{otherwise} . \end{cases}$$

For a time-dependent incompressible velocity field, $v(x,t)$, the Kolmogoroff hypothesis yields the rescaled energy-power spectrum ([9]),

$$(2.6) \qquad \langle |\hat{v}_\delta(k,\omega)|^2 \rangle = \begin{cases} C_0 |k|^{1-d-5/3} \left(|k|^{-2/3} \phi \left(\frac{\omega}{|k|^{2/3}} \right) \right) & , \quad \delta < |k| < 1 \\ 0 & , \quad \text{otherwise} \end{cases}$$

with $\wedge$ the space-time Fourier transform. Here $\phi \geq 0$ is some structure function satisfying $\int \phi(s)ds = 1$. The combination $\omega/|k|^{2/3}$ arises as the only combination of frequency and wave number independent of Re and simultaneously consistent with the scaling in (2.3) and also with the energy spectrum $\int \langle |\hat{v}_\delta(k,\omega)^2\rangle d\omega$ given in (2.5). Thus, the basic problem of turbulent transport in (1.1) with a velocity field with Kolmogoroff statistics can be reformulated as the problem in (2.4) where $v_\delta(x,t)$ has the velocity statistics in (2.6).

Remark: The problem in (2.4) with the velocity statistics in (2.6) assumes for simplicity that there is no fluid motion on the integral length scale. If there is a mean flow varying on the integral length scale, the reformulation in (2.4) is modified to the form,

$$(2.7) \qquad \frac{\partial T}{\partial t} + \bar{V}(\delta x, \delta^{2/3} t) \cdot \nabla T + v_\delta \cdot \nabla T = \kappa_0 \Delta T$$
$$T|_{t=0} = T_0(\delta x) \, , \, \delta \ll 1$$

where v_δ has the velocity statistics in (2.6). The smooth incompressible velocity field $\bar{V}(y, \tau)$ is a given deterministic function and dependence $\bar{V}(\delta x, \delta^{2/3} t)$ indicates that this velocity field varies only on the integral length scale and with the large scale eddy turnover time (see (2.3)) — this mean field represents the non-universal large scale motions while v_δ represents the statistical piece of the turbulent velocity field with arbitrarily many length scales in the limit as $Re \nearrow \infty$, i.e. $\delta \downarrow 0$ (recall that $\delta = (Re)^{-3/4}$). For simplicity in exposition, we assume for the remainder of this section that $\bar{V} \equiv 0$ although some of the phenomena change for $\bar{V} \neq 0$ (see [3] for these results on the model).

With the reformulation in (2.4), here are some of the important theoretical *Goals of an Eddy Diffusivity Theory:*

1) Compute an effective time rescaling function, $\rho^2(\delta)$,

so that the rescaled ensemble average,

$$(2.8)A \qquad \bar{T}(x,t) = \lim_{\delta \downarrow 0} \left\langle T^\delta \left(\frac{x}{\delta}, \frac{t}{\rho^2(\delta)} \right) \right\rangle$$

has a nontrivial limit. (For turbulent transport $\rho^2(\delta)$ should exceed $\delta^{2/3}$, the large scale turnover time).

2) Compute the effective equation satisfied by $\bar{T}(x,t)$ —

$(2.8)B$ this is the "eddy diffusivity" equation because only

large scale fluctuations are involved in this equation (see (2.4)).

Other goals include the computation of higher order moments of the solution, pair dispersion, and properties of interfaces (see [1], [3] for these results for the model).

Remark: We emphasize once again that for turbulent transport, one is interested in results that are uniformly valid independent of Reynolds number as $Re \nearrow \infty$. Since $\delta = (Re)^{-3/4}$, in order to have a theory for turbulent transport which is uniformly valid as $Re \nearrow \infty$, one needs the limit in (2.8)A to exist and be nontrivial.

For the remainder of this section, it will be convenient to consider the parametrized family of incompressible steady velocity fields with energy spectrum given by

$$(2.9)A \qquad \langle |\hat{v}_\delta^{\tilde{\epsilon}}(k)|^2 \rangle = \begin{cases} C_0 |k|^{2-d-\tilde{\epsilon}} & , \quad \delta < |k| < 1 \\ 0 & , \quad \text{otherwise} \end{cases}$$

with $\tilde{\epsilon}, -\infty < \tilde{\epsilon} < \infty$, a parameter — the value of $\tilde{\epsilon} = \frac{8}{3}$ for $d = 3$ yields the spectrum in (2.5). In a similar fashion, we consider a two-parameter family of time-dependent incompressible velocity fields with energy-power spectrum given by

$$(2.9)B \qquad \langle |\hat{v}_\delta(k)|^2 \rangle = \begin{cases} |\hat{v}_\delta^{\tilde{\epsilon}}(k)|^2 \phi \left(\frac{\omega}{|k|^z} \right) & , \quad \delta < |k| < 1 \\ 0 & , \quad \text{otherwise} \end{cases}$$

with $\tilde{\epsilon}$ and z satisfying $0 < z < \infty$ — the values $\tilde{\epsilon} = \frac{8}{3}$ and $z = 2/3$ yield an energy-power spectrum consistent with the Kolmogoroff law in (2.6). For a fixed value of z, the reader can regard $\tilde{\epsilon}$ as a parameter analogous to the $\tilde{\epsilon}$-expansion parameter for renormalization theory from critical phenomena (see [18]). The reason for this terminology will become clear to the reader after the next few paragraphs of discussion.

Homogenization Theory and Mean Field Theory

Basic work of Papanicolaou, Varadhan, and co-workers ([20], [22]) provides a complete theory of eddy diffusivity for the special case of (2.4) involving advection-diffusion by steady incompressible velocity fields provided that there is a sufficient separation of scales between the substantial velocity variations and the initial data, $T_0(\delta x)$. Under these conditions the rigorous theory of eddy diffusivity outlined in (2.8) is the following:

$(2.10)A$ 1) The time-scaling function, $\rho(\delta)$, from (2.8)A is given by the expected diffusive scaling, $\rho(\delta) = \delta$.

2) The eddy diffusivity equation satisfied by

$$\bar{T}(x,t) = \lim_{\delta \downarrow 0} \left\langle T \left(\frac{x}{\delta}, \frac{t}{\delta^2} \right) \right\rangle \text{ is a}$$

simple local diffusion equation,

$(2.10)B$

$$\frac{\partial \bar{T}}{\partial t} = \sum_{i,j=1}^{d} K_{ij}^* \frac{\partial^2 \bar{T}}{\partial x_i \partial x_j} , \quad \bar{T}|_{t=0} = T_0(x)$$

where the formulas for K_{ij}^* are determined from $v(x)$ by solving a stochastic cell problem from homogenization theory (see [20]).

Recently, it has been established ([21], [4]) that the theory in (2.10) remains valid provided that the velocity spectrum of the steady incompressible velocity field in (2.4) satisfies

(2.11)
$$\lim_{\delta \downarrow 0} \int_{R^d} \frac{\langle |\hat{v}_\delta(k)|^2 \rangle}{|k|^2} dk < \infty .$$

The requirement in (2.11) gives a precise condition defining the separation of scales of the significant velocity amplitudes and the initial data in the limit $\delta \downarrow 0$. The condition in (2.11) is satisfied for the steady velocity fields in (2.9)A if and only if $\tilde{\epsilon}$ satisfies $\tilde{\epsilon} < 0$. Thus, the *regime with $\tilde{\epsilon} < 0$ corresponds to mean field theory* — there is separation of scales and the usual diffusive scaling theory for eddy diffusivity outlined in (2.10) applies. In Wilson's theory of critical phenomena, the role of $\tilde{\epsilon}$ was played by the parameter $4-d$ with d, the space dimension; for dimensions $d > 4$ mean field theory applied while for dimensions $d < 4$, new anomalous phenomena occur (see [18]). For transport by steady incompressible velocity fields with the statistics in (2.9)A, a similar role is played by varying the spectral parameter $\tilde{\epsilon}$ in (2.9)A with a fixed space dimension rather than by varying the dimension.

A simple example illustrating the need for the condition in (2.11) is provided by transport-advection with the velocity field given by a periodic simple shear layer, i.e.

(2.12)A
$$\frac{\partial T}{\partial t} + v(x)\frac{\partial T}{\partial y} = \kappa_0 \Delta T , \ T|_{t=0} = T_0(\delta x)$$

with

(2.12)B
$$v(x) = \sum_{j=1}^{\infty} \hat{v}(j)\sin(jx) .$$

It is a simple exercise for the reader to compute the solution of the periodic cell problem mentioned in (2.10)B by separation of variables; the effective diffusion equation is given by

(2.13)A
$$\bar{T}_t = \kappa_0 \Delta \bar{T} + \Delta \kappa \bar{T}_{yy}$$

with

(2.13)B
$$\Delta \kappa = \kappa_0^{-1} \sum_{j=0}^{\infty} \frac{|\hat{v}(j)|^2}{|j|^2}$$

and the diffusive scaling $\rho(\delta) = \delta$ provided that the expression for $\Delta \kappa$ is finite — this example displays the general condition in (2.11) in a transparent fashion. The problems in (2.12) involving simple shear layers were first studied by G.I. Taylor about forty years ago in pioneering work.

Infrared Divergence for the Kolmogoroff Spectrum and the Need for Renormalization

For steady incompressible velocity fields with the Kolmogoroff energy spectrum, $\tilde{\epsilon} = 8/3$, the integral in (2.11) diverges and mean field theory does not occur. In fact

$$(2.14) \qquad \int_{R^d} \langle |\hat{v}_\delta(k)|^2 \rangle dk \to \infty \text{ as } \delta \downarrow 0 \text{ for } \tilde{\epsilon} > 2 ,$$

so there is even an infrared divergence of energy for $\tilde{\epsilon} > 2$. We mention that this fact is not inconsistent with the standard derivation of the Kolmogoroff spectrum which imposes finite energy at the outset. The choice of dissipation space and time scales from (2.3) together with the fact that (2.4) remains uniformly valid in the limit as $Re \nearrow \infty$ results in the infrared divergence of the spectrum *after this rescaling*. Similar divergences occur for turbulent transport by time-dependent velocity fields with the energy power spectra in (2.6) and/or (2.9); however, no simple sharp condition like the one in (2.11) for mean-field theory with the expected diffusive scaling $\rho(\delta) = \delta$ is known at the present time for general time-dependent incompressible velocity fields. Of course there are some important well-known sufficient conditions for mean field theory in the time-dependent case following pioneering ideas of Kubo ([17], [11], [12], [23]). Such infrared divergences indicate that turbulent transport with Kolmogoroff velocity statistics is a difficult problem.

Many authors have attempted to use the methods of field theory and statistical physics to renormalize such divergences in developing theories for eddy diffusivity for turbulent transport (see [16], [27]). Virtually all of these perturbation methods involve formal diagrammatic perturbation theory and partial resummation of divergent perturbation series so an unambiguous assessment of the validity of these methods is needed. One way to achieve this is to develop a class of simple model problems with some of the features of (2.4) and (2.6) where a rigorous theory for renormalized eddy diffusivity satisfying both (2.8)A) and (2.8)B) can be developed. The variety of renormalized perturbation theories can then be checked on such a model. This research program is discussed in the subsequent parts of this paper for the R-N-G theory of Yakhot and Orszag.

2B) A Simple Model with Exact Renormalization for Turbulent Transport. The model problem which is introduced and analyzed by the authors in [1] and [3] is the special case of (2.1) given by

$$(2.15) \qquad \begin{aligned} &\frac{\partial T^\delta}{\partial t} + v_\delta(x,t)\frac{\partial T^\delta}{\partial y} = \kappa_0 \Delta T^\delta \\ &T^\delta|_{t=0} = T_0^\delta(x,y) \equiv T_0(\delta x, \delta y) . \end{aligned}$$

The incompressible velocity field in (2.15) is a simple shearing motion along the y-axis. In basic form, the model problem with a steady velocity coincides with G.I. Taylor's famous example of enhanced diffusion already discussed in (2.13); however, there is one crucial difference —— turbulent velocity statistics for $v_\delta(x,t)$ analogous to those in (2.9) are utilized so that there is a wide range of values of $\tilde{\epsilon}, z$ where infrared divergence occurs (see the discussion above (2.13)), simple mean field theory (see (2.10)) does not apply, and a renormalized theory of eddy diffusivity is needed. In the case of steady velocity fields $v_\delta(x)$ in (2.15), it is assumed in [1] that $v_\delta(x)$ is a stationary Gaussian field with energy spectrum

$$(2.16)A \qquad \left\langle |\hat{v}_\delta^{\tilde{\epsilon}}(k)|^2 \right\rangle = |k|^{1-\tilde{\epsilon}} \psi_0 \left(\frac{|k|}{\delta} \right) \psi_\infty(|k|) .$$

The functions $\psi_0 \left(\frac{|k|}{\delta} \right)$ and $\psi_\infty(|k|)$ are infrared and ultraviolet cut-offs respectively and correspond to the restrictions, $\delta < |k| < 1$ in (2.9), while $\tilde{\epsilon}$ has the same signifi-cance as the spectral parameter in (2.9)A with $d = 1$. For the time dependent case in (2.15), following (2.9)B, the assumption from [1] is that $v_\delta(x,t)$ has stationary Gaussian statistics with energy-power spectrum given by

$$(2.16)B \qquad \left\langle |\hat{v}_\delta(k,\omega)|^2 \right\rangle = \left\langle |\hat{v}_\delta^{\tilde{\epsilon}}(k)|^2 \right\rangle (a|k|)^{-z} \phi \left(\frac{\omega}{a|k|^z} \right)$$

with the specific form $\phi(s) = \pi^{-1}(1 + s^2)^{-1}$; here as in (2.9)B z is a parameter satisfying $0 < z < \infty$. The specific structure function $\phi(s) = \pi^{-1}(1+s^2)^{-1}$ gives the following behavior for the spatial Fourier transform of the temporal auto-correlation function

$$(2.17) \qquad \overbrace{\left\langle v(x,t), v(x,t+t') \right\rangle} = e^{-a|k|^z|t'|} \left\langle |\hat{v}_\delta^{\tilde{\epsilon}}(k)|^2 \right\rangle .$$

Thus, with this structure function, there is the familiar Ornstein-Uhlenbeck expo-nential decay at a given wave number with correlation time given by $(a|k|^z)^{-1}$. In renormalization theories for eddy diffusivity, the interest focuses on long wave numbers (see (2.8)A) with $|k| \ll 1$ thus the behavior in (2.17) yields the following important intuition regarding the role of the parameter z in (2.16)B):

(2.18)
> When z is small there is more rapid
> decorrelation in time at the relevant long wave numbers with
> $|k| \ll 1$ while stronger velocity correlations in time develop at
> these relevant wave numbers as z increases. In particular,
> as $z \nearrow \infty$ the steady velocity statistics in (2.16)A)
> are achieved for $|k| < 1$.

In [1], the rigorous exact renormalization theory for eddy diffusivity with goals outlined in (2.8) is developed for the model problem in (2.15) with the velocity statistics in (2.16) as the parameters $\tilde{\epsilon}$ and z vary. The authors develop a similar renormalization theory (with the same results!) for non-Gaussian statistics in the more recent paper, [33] — thus, there is "statistical universality" for the results presented below. Both the time-rescaling function $\rho(\delta)$ and the effective equation for $\bar{T}(x,t)$ given by the limit

$$(2.19) \qquad \bar{T}(x,t) = \lim_{\delta \downarrow 0} \left\langle T^\delta \left(\frac{x}{\delta}, \frac{t}{\rho^2(\delta)} \right) \right\rangle$$

are determined with complete mathematical rigor as well as limits of higher order moments ([1]), pair dispersion, and the fractal behavior of interfaces ([3]). This rigorous exact renormalization theory is developed through tools involving Fourier analysis and the Feynman-Kac formula and we will not present any details of this analysis here (see [1], [3]). Instead, we describe the remarkable phenomena in the rigorous theory of renormalization that occur for the simple model problem in (2.15) and (2.16) including multiple distinct anomalous scaling regimes as the parameter $\tilde{\epsilon}$ is varied with z fixed and nonlocal equations for eddy diffusivity in some regimes of z and $\tilde{\epsilon}$.

Renormalization Theory for Steady Velocity Fields

In the rigorous renormalization theory for the steady case in (2.15), (2.16)A), there are three distinct scaling regimes for the model problem. The first regime is defined by the parameter range, $\tilde{\epsilon} < 0$. This regime is a region of mean field theory. In the region $\tilde{\epsilon} < 0$, the energy containing length scales associated with the velocity spectrum in (2.16)A) are sufficiently separated from the large scales defined by δ so that (2.8) is satisfied and the classical formulas of homogenization theory apply (see (2.10)); the time scaling law is the expected diffusive scaling, $\rho(\delta) = \delta$; the coefficients for the effective diffusion equation are determined by standard formulas just like the example in (2.13) and depend on the bare viscosity ν_0 as well as the ultraviolet cut-off $\psi_\infty(|k|)$.

The second regime of parameters is determined by $\tilde{\epsilon} > 2$ and corresponds to an anomalous scaling regime dominated by inviscid dynamics. The time scaling function is given by

$$(2.20)A) \qquad \rho(\delta) = \delta^{1-\frac{\tilde{\epsilon}}{4}}, \; 2 < \tilde{\epsilon} < 4 \;,$$

and the effective equation for $\bar{T}$ is a simple diffusion equation

$$(2.20)B) \qquad \frac{\partial \bar{T}}{\partial t} = D(\tilde{\epsilon})t\bar{T}_{yy} \;,$$

where $D(\tilde{\epsilon})$ depends on the ultraviolet cut off $\psi_0(|k|)$ but is independent of the bare diffusivity. Note that the scaling $\rho(\delta) = \delta^{1/2}$ corresponds to the expected convective scaling for the "inviscid" problem in (2.15) with $\nu_0 = 0$ and the time-scales in (2.20)A) are "hyperconvective", i.e. faster than the convective scale. In Section 4 of [1] it is established that the renormalization theory for (2.15) for $\tilde{\epsilon} > 2$ in the steady case coincides with the renormalization theory for the "inviscid" problem.

The third regime of renormalization for the steady case is defined by the inequality, $0 < \tilde{\epsilon} < 2$. The renormalized scaling law

$$(2.21) \qquad \rho(\delta) = \delta^{\frac{1}{1+\frac{\tilde{\epsilon}}{2}}} , \quad 0 < \tilde{\epsilon} < 2$$

is derived in [1]. The function $\rho(\delta)$ in (2.21) agrees with the diffusive scaling at $\tilde{\epsilon} = 0$ but is an anomalous scaling intermediate between the diffusive scaling and the purely convective scaling, $\delta^{1/2}$, which occurs at $\tilde{\epsilon} = 2$. Thus, from (2.20)A) and

(2.21) there are two distinct anomalous scaling regimes for the model problem with steady velocity fields. With the effective equation in (2.20)B for $\tilde{\epsilon} > 2$ and the scaling law in (2.21), naively one might guess that the effective equation for $\bar{T}$ is a simple local diffusion equation with the form,

$$(2.22) \qquad \frac{\partial \bar{T}}{\partial t} = \left(1 + \frac{\tilde{\epsilon}}{2}\right) t^{\frac{\tilde{\epsilon}}{2}} \alpha \mathcal{D} \bar{T}_{yy} \ ,$$

where $\mathcal{D} = (\nu_0)^{\frac{\tilde{\epsilon}}{2}-1}(4\pi)^{-1}$ and α has some prescribed value, say $\bar{\alpha}(\tilde{\epsilon})$, depending on $\tilde{\epsilon}$; the equation in (2.22) has the same form as the equation in (2.20)B for $\tilde{\epsilon} = 2$ and also has the correct scaling with t at $\tilde{\epsilon} = 0$. This naive guess is wrong!! Let $K(y, t, \alpha)$ denote the explicit Green's function for (2.22) for a fixed value of α. This Green's function has the kernel,

$$K(y, t, \alpha) = (4\pi)^{-1/2} t^{-1/2 - \frac{\tilde{\epsilon}}{4}} (\mathcal{D}\alpha)^{-1/2} \exp\left\{\frac{-|y|^2}{4\mathcal{D}\alpha t^{1 + \frac{\tilde{\epsilon}}{2}}}\right\} \ .$$

Then, for each value of $\tilde{\epsilon}$ with $0 < \tilde{\epsilon} < 2$, there is a distribution function of random diffusivities, $\nu_{\tilde{\epsilon}}(\alpha)$, so that the Green's function for the effective equation is given by

$$(2.23)A \qquad K^{\tilde{\epsilon}}(y, t) = \int_{0+}^{\infty} K(y, t, \alpha) d\nu_{\tilde{\epsilon}}(\alpha)$$

with

$$(2.23)B \qquad \bar{T}(x, y, t) = \int K^{\tilde{\epsilon}}(y - \tilde{y}, t) T_0(x, \tilde{y}) d\tilde{y} \ .$$

and $T_0(x, y)$ the (rescaled) initial data. In particular, since $d\nu_{\tilde{\epsilon}}(\alpha)$ is not a point mass, the Green's function from (2.23)A does not have a spatial Gaussian profile for $0 < \tilde{\epsilon} < 2$ and the effective equation for the eddy viscosity theory is not a simple local diffusion equation. The formulas for the distribution function $\nu_{\tilde{\epsilon}}(\alpha)$ for random diffusivity exhibit remarkable changes in complexity as the parameter $\tilde{\epsilon}$ is varied with $0 < \tilde{\epsilon} < 2$ and the interested reader should consult Section 4 of [1] for the details. Some approximate renormalization theories for eddy diffusivity anticipate non-local effective equations [15], [14] but the work in [1] appears to be the first time that explicit formulas for the Green's function have been found without any approximation.

Renormalization Theory for Time Dependent Velocity Fields

The rigorous theory of renormalized eddy diffusivity for the problem in (2.15) with the time dependent velocity statistics in (2.16)B is extremely rich as the parameters $\tilde{\epsilon}$ and z are varied. As indicated in Figure 1, there are five distinct regions in the $\tilde{\epsilon} - z$ upper half-plane with a different structure and scaling law for the rigorous renormalized theory of eddy diffusivity. Next, we give a brief description of this theory of eddy diffusivity for each of the regions depicted in Figure 1. It will be useful for the reader to recall the intuitive significance of the parameter z as described in (2.18).

Region I is defined by the inequalities $\tilde{\epsilon} < 0$ for $z \geq 2$ and $\tilde{\epsilon} < 2 - z$ for $z < 2$ and is a regime of mean field theory while Regions II-V have infrared divergences which require renormalization. The usual diffusive scaling $\rho(\delta) = \delta$ applies in Region I and the effective diffusion equation for $\bar{T}$ is determined by mixed "Kubo-homogenization" formulas with explicit dependence on the bare diffusivity, κ_0 (see the appendix of [1]). It is interesting that for $z \leq 2$, there is sufficiently rapid decorrelation in time in the velocity statistics to push the boundary of mean field theory across $\tilde{\epsilon} = 0$ up to the line $\tilde{\epsilon} = 2 - z$.

Region II is determined by the inequalities, $2 - z < \tilde{\epsilon} < 4 - 2z$ and is a regime with fairly rapid decorrelation in time and weaker infrared divergence. The anomalous scaling function in Region II is given by

$$(2.24)A) \qquad\qquad \rho_2(\delta) = \delta^{\frac{4-\tilde{\epsilon}-z}{2}}$$

and the renormalized diffusion equation for $\bar{T}$ has the form

$$(2.24)B) \qquad\qquad \bar{T}_t = D_2(\tilde{\epsilon}, z)\bar{T}_{yy} \ .$$

The explicit formulas for $D_2(\tilde{\epsilon}, z)$ can be found in [1] but I remark here that they are independent of the bare diffusivity, κ_0.

In Region III defined by the inequalities $4 > \tilde{\epsilon} > 2$ and $\tilde{\epsilon} > 4 - 2z$ there is such a strong singularity in the infrared divergence that the decorrelation in time of the velocity statistics is not important and the renormalization theory coincides exactly with that for the steady case already described in (2.20). Similar remarks describe the renormalization for Region V defined by the inequalities, $z > 2$ and $0 < \tilde{\epsilon} < 2$. Here the values of z are large enough so that the velocity statistics are so strongly correlated that the renormalization theory coincides exactly with that for the steady case described in detail in (2.21) – (2.23). On the boundary with $z = 2$ and $0 < \tilde{\epsilon} < 2$, the same anomalous scaling law as in (2.21) applies but the analogous formulas for the measures, $v_{\tilde{\epsilon}}(\alpha)$, of "random diffusivity" from (2.23) are somewhat different (see Section 5 of [1]).

Finally, we describe the renormalization theory for Region IV defined by the inequalities, $\tilde{\epsilon} < 2$, $z < 2$, and $\tilde{\epsilon} > 4 - 2z$. The anomalous scaling law in this region is given by

$$(2.25)A) \qquad\qquad \rho_4(\delta) = \delta^{\frac{z}{2}\left(\frac{1}{(z-1)+\frac{z}{2}}\right)}$$

and the renormalized diffusion equation has the form

$$(2.25)B) \qquad\qquad \bar{T}_t = t^{1+\frac{z-2}{z}} D_4(\tilde{\epsilon}, z)\bar{T}_{yy}$$

in Region IV; $D_4(\tilde{\epsilon}, z)$ is independent of the bare diffusivity, ν_0.

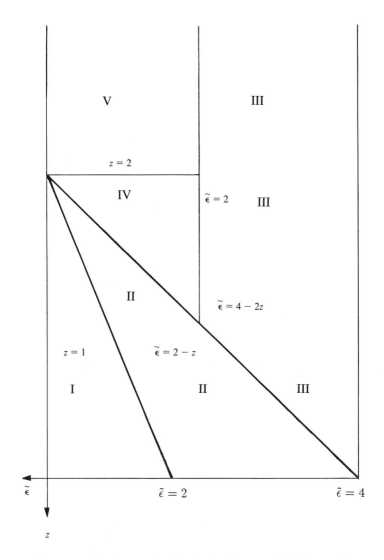

Figure 1: The five regions in the $(\tilde{\epsilon}, z)$ upper half plane with different behavior for the rigorous renormalization theory

Remarks on Renormalization Theory and the Model Problem

Remark 1: As mentioned earlier, the rigorous renormalization theory for the model problem in (2.15) can be developed without the assumption of Gaussian statistics (see [3]). Furthermore, for steady velocity fields, there is a second completely different approach involving renormalization of Stieltjes measure formulas for eddy diffusivity ([4], [5]) which has been developed by the authors in other work; the renormalization theory for (2.15) for the case with velocity fields with almost periodic coefficients is developed using this approach in [6].

Remark 2: With the identification from (2.6) and (2.9) in the velocity statistics in (2.16), the analogue of the Kolmogoroff spectrum in the model problem occurs at the values $\tilde{\epsilon} = \frac{8}{3}$ and $z = \frac{2}{3}$ — this is a value at the boundary between Region II and Region III. The effective diffusivity exhibits an extremely interesting behavior at this boundary in the model problem (see Section 5 of [1]).

Remark 3: Clearly the simple model problem does not incorporate all the features of general turbulent transport in R^d; nevertheless, one might hope to gain insight into these general problems in R^d through the behavior of the model. Very recently the authors ([7]) have shown that analogous scalings with a more complex theory for eddy diffusivity occur in R^d with the velocity statistics in (2.9) in the analogous regime to Region III in Figure 2 with the *same anomalous scaling law*, $\rho(\delta) = \delta^{1-\frac{\tilde{\epsilon}}{4}}$ *as given for the model problem.* By other arguments, the authors have also established that there is analogous renormalization for Region II as in (2.24)A),B) in the model for general turbulent transport in R^d. Thus, the authors have utilized the model to understand the behavior of renormalized theories of eddy diffusivity for turbulent transport in the vicinity of the Kolmogorov spectrum!! All of these results are summarized in a recent physics letter of the authors ([8]).

Finally, we mention that the theory of renormalized eddy diffusivity for the model problem with non-zero mean fields such as those described in (2.7) is developed by the authors in [3] — some of the regions of renormalization, specifically IV and V, are very sensitive to such effects of non-trivial mean fields while Region III and II (for $z < 1$) are completely insensitive to such mean fields (see [3] for the details).

The rigorous renormalization theory for the model problem in (2.15) provides an unambiguous test problem for the many approximate renormalization theories for eddy diffusivity that have been developed through ad hoc techniques or formal analogies with statistical physics ([16], [25], [10], [24], [14], [27]). An excellent survey of many of these approximate renormalization theories for eddy diffusivity for turbulent transport is given in the recent paper by Kraichnan ([15]). In the remainder of this paper, we illustrate the importance of having a rigorous non-trivial exactly solvable model by comparing the predictions of the recent renormalization theory due to Yakhot and Orszag ([26], [27]) with the exact renormalization theory described in Section 2B) (see [1]). More comparisons between various approximate and exact renormalization theories for the model problem are presented in [2].

SECTION 3: THE R-N-G METHOD APPLIED TO THE SIMPLIFIED MODEL

The R-N-G method of Yakhot-Orszag ([26]) follows the overall strategy of renormalization methods for turbulence pioneered by Rose ([25]) and Foster-Nelson-Stephen ([10]) in a fashion which differs in the details and facilitates explicit computation of renormalized coefficients. In turn the work in [25] and [10] is based on adapting renormalization ideas from the modern theory of critical phenomena

([18]) to problems in turbulence. This analogy is naturally exploited in Fourier space and the starting point is to take the space-time Fourier transform of the dynamic equations. The space-time Fourier transform of the model equation in (2.15) is the equivalent *Integral Equation in Fourier Space*

$$(2\pi i\omega + 4\pi^2\kappa_0|k|^2)\hat{T}^\delta(k,\omega)$$

(3.1)

$$= -2\pi i k_2 \int\int \hat{v}_\delta(k_1 - k_1', \omega - \omega')\hat{T}^\delta(k_1', k_2, \omega')dk_1'd\omega' + \hat{T}_0^\delta(k_1, k_2) \ .$$

Here we use the Fourier transform convention,

(3.2)
$$\hat{f}(k,\omega) = \int\int\int e^{-2\pi i(xk_1 + yk_2 + \omega t)} f(x,y,t)dxdydt$$

and we assume for simplicity in exposition that the initial data in (2.15), $T_0^\delta(x,y) \equiv T_0(\delta x, \delta y)$, is defined through a function $T_0(x,y)$ so that its Fourier transform,

(3.3)
$$\hat{T}_0(k_1, k_2) \text{ vanishes for } |k| > R_0 > 0 \ .$$

Before beginning the discussion of renormalization, we introduce the precise class of velocity statistics satisfying (2.16) and (2.17) which we use here that are particularly natural for discussing an R-N-G method.

Velocity Statistics

In the case of steady velocity fields, we assume that the random velocity field $v_\delta^{\bar{e}}(x)$ is given by

(3.4)
$$v_\delta^{\bar{e}}(x) = \int_{\delta < |k| \le \Lambda_0} e^{2\pi i(k \cdot x)} |k|^{\frac{1-\bar{e}}{2}} W(dk)$$

where $W(dk)$ is Gaussian white noise satisfying $\langle W(dk)\overline{W(dk')}\rangle = \delta(k + k')dk$. The velocity field in (3.4) satisfies (2.16)A) with the special infrared and ultraviolet cut-offs given by

(3.5)
$$\psi_0(|k|) = \begin{cases} 1, & |k| > \delta \\ 0, & |k| < \delta \\ \text{and} \\ \psi_0(|k|) = \begin{cases} 1, & |k| < \Lambda_0 \\ 0, & |k| > \Lambda_0 \ . \end{cases} \end{cases}$$

For the general space time velocity field, $v_\delta(x,t)$, we assume the form

(3.6) $v_\delta(x,t) = \pi^{-1/2}\int_\delta^{\Lambda_0}\int_{-\infty}^{\infty} e^{2\pi i(kx + a|k|^z\omega t)}|k|^{\frac{1-\bar{e}}{2}}(1 + \omega^2)^{-1/2}W(d\omega)\otimes W(dk)$

where $W(d\omega)\otimes W(dk)$ is two-dimensional Gaussian white noise satisfying $\langle W(d\omega)\otimes W(dk), W(d\omega')\otimes W(dk')\rangle = \delta(k + k')\delta(\omega + \omega')dkd\omega$; from this orthogonality relation, it follows that the correlation function for the velocity field is given by

$$\langle v_\delta(x + x', t + t')v_\delta(x,t)\rangle =$$

(3.7)

$$\int_{\delta < |k| < \Lambda_0} e^{2\pi i k x'} e^{-2\pi a|k|^z|t'|} |k|^{1-\bar{e}} dk$$

so that in particular (2.17) is satisfied. We remark that the calculation of (3.7) from (3.6) relies on the well-known calculus formula

$$\pi^{-1} \int e^{i\alpha s}(1+s^2)^{-1}ds = e^{-|\alpha|} \ .$$

The velocity fields used here are a special case those defined in [1] with special cut-offs which are especially convenient for renormalization theory; we have taken special care in defining the velocity fields used here in order to keep track of explicit constants in comparing the exact and approximate R-N-G theories.

3A) Finite Band Mode Elimination in R-N-G Methods. The starting point for R-N-G methods is decomposition of the integral equation in (3.1) into modes involving long wavelengths and short wavelengths and then to systematically eliminate the short wavelength contributions. We set $\Lambda(r) = e^{-r}\Lambda_0$ where r is a parameter with $0 < r < +\infty$ and introduce the characteristic function, $\chi^{r,\tilde{r}}(k)$, defined by

(3.8)
$$\chi^{r,\tilde{r}}(k) = \begin{cases} 1, & |k_1| \le e^{-\tilde{r}}\Lambda(r) \\ 0, & \text{otherwise} \end{cases}$$

where $\tilde{r}$ with $0 < \tilde{r} < +\infty$ is a second parameter. For the moment we set $r = 0$ in (3.8) and define lower mode components via

(3.9)A)
$$\hat{v}^<(k,\omega) = \chi^{0,\tilde{r}}(k)\hat{v}(k,\omega)$$
$$\hat{T}^<(k,\omega) = \chi^{0,\tilde{r}}(k)\hat{T}(k,\omega)$$

with $\hat{v}^>(k,\omega), \hat{T}^>(k,\omega)$ defined by

(3.9)B)
$$\hat{v}^>(k,\omega) = \hat{v}(k,\omega) - \hat{v}^<(k,\omega)$$
$$\hat{T}^>(k,\omega) = \hat{T}(k,\omega) - \hat{T}^<(k,\omega) \ .$$

After dividing the integral equation in (3.1) by the Fourier transform of the free space Green's function and using the decomposition in (3.9), the integral equation in (3.1) can be rewritten in the schematic form

(3.10)
$$A) \ \hat{T}^<(k,\omega) = \lambda A^<\hat{T}^< + \lambda B^<\hat{T}^> + C^<\hat{T}_0$$
$$B) \ \hat{T}^>(k,\omega) = \lambda A^>\hat{T}^> + \lambda B^>\hat{T}^<$$

where we will not write down the explicit form of these operators at this stage in the discussion; the parameter λ in (3.10) is a ficticious coupling constant which is assumed to be small in the perturbation argument sketched below — in fact, in the real equations in (3.1) λ is always evaluated at the value, $\lambda = 1$. Since the goal in theories of eddy diffusivity is to understand the large scale motions, the first step in an R-N-G theory is to eliminate the effects of $\hat{T}^>$ in (3.10)A) by an approximate solution of (3.10)B). Assuming $\lambda \ll 1$ in (3.10), utilizing a Neumann series expansion, and inserting the result in (3.10)A), we obtain,

(3.11)
$$\hat{T}^<(k,\omega) = \lambda A^<\hat{T}^< + \lambda^2 B^<B^>\hat{T}^< + C^<\hat{T}_0^\delta + 0(\lambda^3) \ .$$

Neglecting the terms $0(\lambda^3)$ formally and setting $\lambda = 1$, the result is the approximate equation for $\hat{T}^<(k,\omega)$ alone,

$$(3.12) \qquad \hat{T}^<(k,\omega) = A^<\hat{T}^< + B^<B^>\hat{T}^< + C^<\hat{T}_0^\delta$$

where the short wavelength modes in $\hat{T}(k,\omega)$ have all been eliminated in (3.12) — this approximation in (3.12) is the *first order mode elimination scheme*.

The operators $A^<$ and $B^<B^>$ involved both $v^>$ and also $v^<$; the short wavelength contributions from $v^>$ to these operators are eliminated from (3.11) by *statistical averaging* over the modes with $|k_1| > e^{-\tilde{r}}\Lambda_0$. The goal of the finite band elimination scheme in R-N-G theories is to show through this process that the equation from (3.12) for $\hat{T}^<(k,\omega)$ alone resembles the original equation in (3.1) except that there is an augmented viscosity through this higher mode averaging procedure. This *augmented viscosity can then be used through successive iterations of the same mode elimination scheme to balance the strong infrared divergences* when needed — this basic observation of the potential use of R-N-G in this context is originally due to Rose ([25]). Nevertheless, even with averaging over the velocity statistics, the calculation of $B^<B^>\hat{T}^<$ is difficult to achieve exactly in a fashion which readily displays the effects of enhancing the diffusivity. In the R-N-G theory of Yakhot and Orszag ([26], [27]), calculational simplicity is achieved through what we call a *Principle of Distant Interaction* which is ubiquitous throughout the calculations from [26], [27]; we postpone a discussion of this principle until the appendix where it is discussed in detail for the exactly solvable model problem from (2.15). Thus, the basic finite band R-N-G procedure of Yakhot and Orszag relies on three approximations:

(3.13)
1) *First Order High Wave Number Mode Elimination*
   (see (3.11) and (3.12) above)
2) *Partial Averaging of High Wave Number Velocity Statistics*
3) *Principle of Distant Interaction.*

For the simple model problem in (2.15), a systematic application of these approximation procedures in (3.13) yields the

Equation with Enhanced Diffusivity after Finite Mode Elimination:
Eliminating the modes $\Lambda_0 e^{-\tilde{r}} \leq |k_1| \leq \Lambda_0$, we have

$$(3.14)A \qquad \begin{aligned} &(2\pi i\omega + 4\pi^2\kappa_0 k_1^2 + 4\pi^2(\kappa_0 + \Delta\kappa(\tilde{r}))k_2^2)\hat{T}^<(k,\omega) \\ &= -2\pi i k_2 \int\int \hat{v}^<(k_1 - k_1', \omega - \omega')\hat{T}^<(k_1', k_2, \omega')dk_1'd\omega' + \hat{T}_0^\delta(k) \end{aligned}$$

where the augmented viscosity, $\Delta\kappa(\tilde{r})$, is given by

$$(3.14)B \qquad \Delta\kappa(\tilde{r}) = 2\int_{e^{-\tilde{r}}\Lambda_0}^{\Lambda_0} (2\pi a|k|^z + 4\pi^2\kappa_0|k|^2)^{-1}|k|^{1-\tilde{\epsilon}}dk$$

provided $e^{-\tilde{r}}\Lambda_0 \geq \delta$. For the steady velocity fields in (3.4), we use (3.14)B with $a = 0$. Clearly the formula for the increased diffusivity, $\Delta\kappa(\tilde{r})$, in (3.14)B depends on the parameters $\tilde{\epsilon}, z$ defining the velocity statistics.

As we mentioned earlier, in situations with infrared divergence, the whole point of R-N-G is to successively apply a basic mode elimination scheme in an iteration procedure; thus, the equations in (3.14) are really a special case of the following

General Equations with Enhanced Diffusivity after Finite Mode Elimination

Assume that $\hat{T}_r(k,\omega)$ involves only modes with

$|k_1| \leq \Lambda_0 e^{-r}, i.e. \chi^{0,r}\hat{T}_r = \hat{T}_r$ and satisfies the integral equation,

$(2\pi i\omega + 4\pi^2\kappa_0 k_1^2 + 4\pi^2\kappa(r)k_2^2)\hat{T}_r$

(3.15)
$$= -2\pi i k_2 \int\int \hat{v}_r(k_1 - \tilde{k}, \omega - \tilde{\omega})\hat{T}_r(\tilde{k}, k_2, \tilde{\omega})d\tilde{k}d\tilde{\omega} + \hat{T}_0^\delta$$

where $\hat{v}_r = \chi^{0,r}\hat{v}$ and $\kappa(r)$ is a general viscosity coefficient.

If the modes k_1, for $e^{-\tilde{r}}\Lambda(r) \leq |k_1| \leq \Lambda(r)$ are eliminated with $e^{-\tilde{r}}\Lambda(r) \geq \delta$ using the approximations in (3.13)

then $\hat{T}^< = \chi^{r,\tilde{r}}\hat{T}_r$ satisfies the augmented viscosity equation

$(2\pi i\omega + 4\pi^2\kappa_0 k_1^2 + 4\pi^2(\kappa(r) + \Delta\kappa(r,\tilde{r})k_2^2)\hat{T}^<$

(3.16)A
$$= -2\pi i k_2 \int\int \hat{v}(k_1 - \tilde{k}, \omega - \omega')\hat{T}^<(\tilde{k}, k_2, \tilde{\omega})d\tilde{k}d\tilde{\omega} + \hat{T}_0^\delta$$

where $\hat{v}^< = \chi^{r,\tilde{r}}\hat{v}$ and

(3.16)B $\quad \Delta\kappa(r,\tilde{r}) = 2\int_{e^{-\tilde{r}}\Lambda}^{\Lambda}(2\pi a|k|^z + 4\pi^2\kappa_0|k|^2)^{-1}|k|^{1-\tilde{\epsilon}}dk$.

For the special case of steady velocity fields the formula in (3.16)B applies with $a = 0$.

Of course (3.14) is a special case of (3.16) with $\Lambda(r) = \Lambda_0$ and $\kappa(r) = \kappa_0$. All the technical details utilizing the principles in (3.13) to derive the equations in (3.16) are discussed in the appendix to this paper; in particular, we provide a detailed discussion of the Principle of Distant Interaction. The mode elimination scheme in (3.15), (3.16) is the main technical device for the R-N-G methods developed here.

As mentioned earlier, there are other similar R-N-G procedures for turbulent transport in the literature (see [19] and [25]); they all seem to be based on the same philosophy of mode elimination but differ greatly in the technical details in Steps 1) and 3) from (3.13) — it would be interesting to check the performance of some of these alternative R-N-G procedures on the model problem from (2.15).

3B) Finite Mode Elimination, Mean Field Theory, and Infrared Divergence. If we attempt to eliminate all the high wave number modes in a single step, we set $\delta = e^{-\tilde{r}}\Lambda_0$ in (3.14)B and obtain that the augmented viscosity is given by

(3.17)
$$\Delta\kappa^\delta = 2\int_\delta^{\Lambda_0}(2\pi a|k|^z + 4\pi^2\kappa_0|k|^2)^{-1}|k|^{1-\tilde{\epsilon}}dk .$$

Thus, we immediately obtain the following facts:

(3.18)A) The additional diffusivity is finite in the limit as

$\delta \downarrow 0$ if and only if $\tilde{\epsilon} < \max\{0, 2 - z\}$

and is given by $\Delta\kappa = \lim\limits_{\delta\downarrow 0} \Delta\kappa^\delta =$

$$2 \int_0^{\Lambda_0} (2\pi a|k|^z + 4\pi^2 \kappa_0 |k|^2)^{-1} |k|^{1-\tilde{\epsilon}} dk .$$

Thus, $\tilde{\epsilon} < \max\{0, 2 - z\}$ defines the regime of mean field

theory predicted by the mode elimination procedure. There is

(3.18)B) *infrared divergence of effective diffusivity provided that*

$\tilde{\epsilon} > \max\{0, 2 - z\}$ in the sense that

$\lim\limits_{\delta\downarrow 0} \Delta\kappa^\delta = +\infty$ for $\tilde{\epsilon} > \max\{0, 2 - z\}$.

Next we compute the large scale long-time limit equation satisfied by $\bar{T}(x, t) = \lim_{\delta\downarrow 0} \left\langle T^\delta \left(\frac{x}{\delta}, \frac{t}{\rho^2(\delta)} \right) \right\rangle$ in the asymptotic limit procedure by rescaling (3.14)A) in the limit $\delta \downarrow 0$. In general we note that

with the change of variables

(3.19) $x' = \delta x, \ y' = \delta y, \ t' = \rho^2(\delta)t$

the dual variables with $k = (k_1, k_2)$ change according to

$k' = \delta^{-1}k, \ \omega' = (\rho^2(\delta))^{-1}\omega$

so that finding the effective equation satisfied by

$$\bar{T}(ax', y', t') = \lim_{\delta\to o} \left\langle T^\delta \left(\frac{x'}{\delta}, \frac{y'}{\delta}, \frac{t'}{\rho^2(\delta)} \right) \right\rangle$$

(3.19)B) is equivalent through Fourier transform to computing

$$\hat{\bar{T}}(k', \omega') = \lim_{\delta\to 0} \left\langle \hat{T}^\delta(\delta k', \rho^2(\delta)\omega') \right\rangle .$$

To compute the diffusion equation satisfied by $\bar{T}$ as predicted by the finite mode elimination scheme in the mean field regime, $\tilde{\epsilon} < \max\{0, 2 - z\}$, we set $\delta = e^{-r}\Lambda_0$ in (3.14)A) so that in particular $v^< \equiv 0$ and set $\rho(\delta) = \delta$, the usual diffusion scaling; then the limit of (3.14)A) becomes the equation

(3.20)
$$(2\pi i\omega' + \kappa_0 4\pi^2 (k_1')^2 + 4\pi^2 (\kappa_0 + \Delta\kappa)(k_2')^2)\hat{\bar{T}}(k', \omega')$$
$$= \hat{T}_0(k') .$$

Therefore, the equation in (3.20) implies that the finite mode elimination scheme described in Section 3A) predicts the eddy diffusivity equation for $\bar{T}$ given by

(3.21)
$$\hat{\bar{T}}_t = \kappa_0 \Delta\bar{T} + \Delta\kappa\bar{T}_{yy}$$
$$\bar{T}|_{t=0} = T_0(x, y)$$

with the diffusion scaling $\rho(\delta) = \delta$ and $\Delta\kappa$ given in (3.18)A) provided that $\tilde{\epsilon} < \max\{0, 2 - z\}$.

Remark: The *predictions of the mean field regime* using the finite mode elimination scheme *are exact for the model problem in (2.15).* The mean field region (see Region I) as well as the coefficient formulas for the eddy diffusivity are reproduced exactly! This is a special feature of the model problem in (2.15); in general, the first order perturbation expansion used in Step (1) from (3.13) is typically not exact (see [4]) and additional errors occur if the perturbation scheme in (3.13)A is used for general turbulent transport in mean field regimes. In fact, the first order mode elimination always computes the upper bound for enhanced diffusivity (see [4]) in the steady case and this upper bound is achieved by simple shear layers; this explains why the eddy diffusivity equation is exact in (3.21).

3C) Infinitesmal Mode Elimination and the Renormalization of Infrared Divergences.

Here we consider the application of the R-N-G method in the regime $\tilde{\epsilon} > \max\{0, 2 - z\}$ where there is infrared divergence in the straightforward single step mode elimination scheme presented in 3B. The basic overall strategy in R-N-G is to eliminate small bands successively so that the increase in diffusivity (from (3.16)B in our application) can be used to compensate the infrared divergence (described in (3.18)B in our application). As regards the goals of a theory for eddy diffusivity as mentioned in (2.8), the R-N-G procedure will predict the nonlinear scaling function $\rho(\delta)$ and the "eddy diffusivity" equation which emerges with this predicted scaling as the *renormalized fixed point equation* in the R-N-G terminology from critical phenomena ([18], [10], [26]). In the simple example of mean field theory, we have already illustrated a simple version of this procedure in (3.20) yielding (3.21); hopefully, the terminology introduced above will become clear to the reader after we discuss this below.

In R-N-G methods, either a finite difference scheme ([25]) is utilized for iterating the effects on viscosity of successive mode elimination of small bands or a differential equation ([10], [26]) is utilized for assessing the augmented viscosity through mode elimination of an infinitesmal band. Next following Yakhot-Orszag ([26]) we derive a differential equation for enhanced diffusivity through R-N-G in the model problem and then solve this differential relation as $r \to \infty$.

The Differential Equation for Enhanced Diffusivity

After the band of modes $\Lambda(r) \leq |k_1| \leq \Lambda_0$ have been eliminated through R-N-G resulting in an enhanced diffusion coefficient $\kappa(r)$ (see (3.16)A above), we want to derive a differential equation for $\kappa(r)$ which results from eliminating the finite band of modes, $e^{-\tilde{r}}\Lambda(r) \leq |k_1| \leq \Lambda(r)$, in the infinitesmal limit as $\tilde{r} \to 0$. We utilize (3.16)B and compute that

(3.22)
$$\frac{d\kappa(r)}{dr} = \lim_{\tilde{r} \to 0} \frac{\Delta\kappa(r,\tilde{r})}{\tilde{r}} =$$
$$2 \lim_{\tilde{r} \to 0} \frac{\Lambda(r)(1 - e^{-\tilde{r}})}{\tilde{r}} \lim_{\tilde{r} \to 0} (\Lambda(r)(1 - e^{-\tilde{r}}))^{-1} \int_{e^{-\tilde{r}}\Lambda(r)}^{\Lambda(r)} (2\pi a|k|^z + 4\pi^2\kappa_0|k|^2)^{-1}|k|^{1-\tilde{\epsilon}}dk \ .$$

Clearly, the limit in (3.22) results in the following *Differential Equation for En-*

hanced Diffusivity:

$$(3.23) \qquad \begin{aligned} \frac{d\kappa(r)}{dr} &= 2\Lambda(r)^{2-\tilde{\epsilon}} \left[2\pi a(\Lambda(r))^z + 4\pi^2 \kappa_0 (\Lambda(r))^2 \right]^{-1} \\ \kappa(r)|_{r=0} &= \kappa_0 , \qquad\qquad\qquad \Lambda(r) = e^{-r}\Lambda_0 . \end{aligned}$$

Since we are interested in the total enhanced diffusivity through elimination of all modes, $\delta \le |k_1| \le \Lambda_0$, by iterating the "infinitesmal" R-N-G mode elimination procedure, we set $\delta = e^{-r}\Lambda_0$ with $\delta \ll 1$ and focus on the asymptotic behavior of the solution of (3.23) in the limit $r \to \infty$. This elementary problem is easily solved with an interesting role for the two parameters, $\tilde{\epsilon}$ and z, measuring the strength of infrared divergence and amount of decorrelation at long wave lengths.

We have the following

PROPOSITION 1: *The solution $\kappa(r)$ of the O.D.E. in (3.23) has the following asymptotic behavior as $r \to \infty$ for $\tilde{\epsilon} > \max\{0, 2 - z\}$:*

A) For steady velocity fields and $-\infty < \tilde{\epsilon} < \infty$,

$$\kappa(r) = \kappa_0 + \frac{2\Lambda_0^{-\tilde{\epsilon}}}{4\pi^2 \kappa_0} \frac{(e^{\tilde{\epsilon}r} - 1)}{\tilde{\epsilon}} .$$

B) For $z > 2$ and $\tilde{\epsilon} > 0$, the asymptotic behavior is exactly like that in the steady case from A), i.e.

$$\kappa(r) = \frac{2}{4\pi^2 \kappa_0} \Lambda_0^{-\tilde{\epsilon}} \tilde{\epsilon}^{-1} e^{\tilde{\epsilon}r} (1 + 0(e^{(2-z)r})) \text{ as } r \to \infty .$$

(3.24)

C) For $z = 2$ and $\tilde{\epsilon} > 0$

$$\kappa(r) = \kappa_0 + \frac{2\Lambda_0^{-\tilde{\epsilon}} \tilde{\epsilon}^{-1}}{2\pi a + 4\pi^2 \kappa_0} (e^{\tilde{\epsilon}r} - 1) .$$

D) For $z < 2$ and $\tilde{\epsilon} > 2 - z$

$$\kappa(r) = (\pi a)^{-1} \frac{\Lambda_0^{2-\tilde{\epsilon}-z}}{(\tilde{\epsilon} - 2 + z)} e^{(\tilde{\epsilon}-2+z)r} (1 + 0(e^{(z-2)r}))$$

as $r \to \infty$.

From the above proposition, we see that the *enhanced diffusivity* from the *iterated R-N-G increases exponentially with r* for values of $\tilde{\epsilon}, z$ which satisfy $\tilde{\epsilon} > \max\{0, 2 - z\}$, i.e. in the complement of mean-field theory. In contrast this contribution from long wave numbers to the enhanced diffusivity decreases exponentially in the mean field regime with $\tilde{\epsilon} < \max\{0, 2 - z\}$; for example, in the steady case, we obtain from A) of the proposition that

$$\lim_{r \to \infty} \kappa(r) = \kappa_0 - \frac{2\Lambda_0^{-\tilde{\epsilon}}}{4\pi^2 \kappa_0 \tilde{\epsilon}} \text{ for } \tilde{\epsilon} < 0 ,$$

and this is the exact value for augmented diffusivity computed in (3.18)A) by the one-step mode elimination procedure. This last comment provides some further justification for the use of the infinitesmal iterated R-N-G procedures described above. The calculation of the asymptotic behavior of $\kappa(r)$ plays a crucial role in the R-N-G computation of an eddy diffusivity. We discuss this next.

Anomalous Scaling Exponents and the Renormalized Fixed Point Equation

To summarize the developments of the R-N-G procedure to this point, we have established that through the infinitesmal mode elimination algorithm, $\hat{T}_r(k, \omega)$, involving only modes with $|k_1| \leq \Lambda(r)$, i.e. $\chi^{0,r}\hat{T}_r = \hat{T}_r$, satisfies the equation in (3.15) (which we reproduce here),

$$
\begin{aligned}
(3.25) \quad & (2\pi i\omega + 4\pi^2 \kappa_0 k_1^2 + 4\pi^2 \kappa(r)k_x^2)\hat{T}_r(k, \omega) \\
& = -2\pi i k_1 \int\int \hat{v}_r(k_1 - \tilde{k}, \omega - \tilde{\omega})\hat{T}_r(\tilde{k}, k_2, \tilde{\omega})d\tilde{k}d\tilde{\omega} + \hat{T}_0^\delta
\end{aligned}
$$

provided that $\Lambda(r) \geq \delta$ with $\Lambda(r) = e^{-r}\Lambda_0$. Here the coefficient of enhanced diffusivity, $\kappa(r)$, satisfies the O.D.E. in (3.23) with the asymptotic properties given in (3.24) from Proposition 1 as $r \to \infty$ for $\tilde{\epsilon}, z$ satisfying $\tilde{\epsilon} > \max\{0, 2 - z\}$.

To compute the anomalous scaling exponents and the eddy diffusivity equation from (2.8), the R-N-G theory involves the computation of the renormalized fixed point equation ([18], [26]); thus, the equation in (3.25) is rescaled to unit size in wave-number so that the equation in (3.25) converges to a nontrivial limit equation as $r \to \infty$. This equation is the renormalized fixed point equation and if we recall the correspondence in (3.19), this equation should be the Fourier transform of the eddy diffusivity equation satisfied by $\bar{T}$ as predicted by the R-N-G theory; the anomalous scaling exponents are determined uniquely by the requirement that this renormalized fixed point equation is non-trivial.

Since both $\hat{T}_r$ and $\hat{v}_r$ as well as $\hat{T}_0^\delta$ have their support in $|k_1| \leq \Lambda(r)$ for $\Lambda(r) \geq \delta$, we introduce the rescalings

$$
\begin{aligned}
(3.26)A \quad & k' = e^r \Lambda_0 k \\
& \omega' = e^{2\alpha r}\Lambda_0^{2\alpha}\omega
\end{aligned}
$$

with

$$
\begin{aligned}
(3.26)B \quad & \hat{T}_r(k', \omega') = \beta_r \hat{T}_r(e^{-r}\Lambda_0^{-1}k', e^{-2\alpha r}\Lambda_0^{2\alpha}\omega') \\
& v_r(k'\omega') = \hat{v}_r(e^{-r}\Lambda_0^{-1}k', e^{-2\alpha r}\Lambda_0^{-2\alpha}\omega') \\
& \beta_r = e^{-(2\alpha+2)r}\Lambda_0^{-(2\alpha+2)} .
\end{aligned}
$$

The coefficient α in (3.26)A is the anomalous scaling exponent and remains to be determined as part of the R-N-G procedure while the coefficient β_r is introduced to balance the Fourier transform of the initial data (since the problem is linear, β_r is introduced here mainly for aesthetic reasons). We make the important remark that with the identification from (3.19)A, the anomalous scaling exponent from (2.8), $\rho(\delta)$ is given by

$$
(3.27) \quad \rho(\delta) = \delta^\alpha \text{ provided that } \delta = e^{-r}\Lambda_0 .
$$

Through straightforward calculation utilizing (3.26), we compute that equation (3.25) in the rescaled variables becomes

$$2\pi i\omega' + e^{(2\alpha-2)r}\Lambda_0^{2\alpha-2}4\pi^2\kappa_0(k_1')^2 + 4\pi^2\kappa(r)e^{(2\alpha-2)r}\Lambda_0^{2\alpha-2}(k_2')^2)\hat{T}_r(k',\omega')$$

$$(3.28) \quad = -e^{-2r}\Lambda_0^{-2}2\pi i k_2' \int\int \tilde{v}_r(k'-\tilde{k}',\omega'-\tilde{\omega}')\hat{T}_r(\tilde{k}',\tilde{\omega}')d\tilde{k}'d\tilde{\omega}'$$

$$+ \delta^{-2}e^{-2r}\Lambda_0^{-2}\hat{T}_0(\delta e^{-r}\Lambda_0^{-1}k') .$$

Since the equation in (3.25) remains valid for $\Lambda(r) \geq \delta$, we set $\Lambda_0 e^{-r} = \delta$ in (3.28). In the large scale limit, $\delta \ll 1$, necessarily $r \to \infty$ so the asymptotic formulas for $\kappa(r)$ from (3.24) in Proposition 1 apply and $\kappa(r) \to \infty$ as $\alpha \downarrow 0$ for $\tilde{\epsilon} > \max\{0, 2-z\}$; thus, there is a unique power of α for each $\tilde{\epsilon}, z$ so that $\kappa(r)e^{(2\alpha-2)r}$ remains finite in the limit. By (3.24) of Proposition 1

A) for steady velocities or unsteady velocities with $\tilde{\epsilon} > 0, z > 2$,

$$\alpha = 1 - \frac{\tilde{\epsilon}}{2} \quad \text{and}$$

$$(3.29) \qquad \lim_{r\to\infty} \kappa(r)e^{(2\alpha-2)r}\Lambda_0^{2\alpha-2} = \frac{2}{4\pi^2\kappa_0}\tilde{\epsilon}^{-1} \equiv D_A(\tilde{\epsilon}) .$$

B) For unsteady velocities with $z < 2$ and $\tilde{\epsilon} > 2 - z$

$$\alpha = \frac{4-(\tilde{\epsilon}+z)}{2} \quad \text{and}$$

$$\lim_{r\to\infty} \kappa(r)e^{(2\alpha-2)r}\Lambda_0^{2\alpha-2} = (\pi a)^{-1}(\tilde{\epsilon}-2+z)^{-1} \equiv D_B(\tilde{\epsilon}, z) .$$

We leave the case $z = 2$ and $\tilde{\epsilon} > 0$ as an easy exercise for the reader. Thus, from (3.27) we compute that the anomalous scaling exponents predicted by R-N-G are given by

$$(3.30) \quad \begin{aligned} &A) \; \rho_A(\delta) = \delta^{1-\frac{\tilde{\epsilon}}{2}}, &&\tilde{\epsilon} > 0, z \geq 2 \text{ for the steady velocities} \\ &B) \; \rho_B(\delta) = \delta^{\frac{4-(\tilde{\epsilon}+z)}{2}}, &&\tilde{\epsilon} > 2 - z \text{ and } z < 2 . \end{aligned}$$

We also observe that for any values of $\tilde{\epsilon}, z$ with $\tilde{\epsilon} > \max\{0, 2 - z\}$, the anomalous exponent, α, satisfies $\alpha < 1$ so that the effects of molecular diffusivity, κ_0 vanish in the limit because $e^{(2\alpha-2)r}\omega \to 0$ as $r \to \infty$. Furthermore, with $e^{-r}\Lambda_0 = \delta, \hat{v}_r \equiv 0$ because we have the special infrared cut-off in (3.5), (3.6). The perceptive reader will note that the argument we have just presented is actually valid only for initial data T_0 with the support of $\hat{T}$ contained in the unit ball. For more general data with $\hat{T}$ of compact support and for more general infrared cut-offs this simple argument would not work, however we can still show that the contribution from the renormalized convolution on the right hand side of (3.28) is negligible. Here is the reasoning — by definition $\tilde{v}_r(k',\omega')$ has support confined to $1 \leq |k'| \leq \Lambda_0$ so this velocity defining the convolution kernel is both infrared and ultraviolet convergent; furthermore, this term is multiplied by $e^{-2r}\Lambda_0^{-2} = \delta^2$ and therefore vanishes in the limit as $\delta \downarrow 0$. To summarize, the *renormalized fixed point equation* as $\delta \downarrow 0$ is given by

$$(3.31) \qquad (2\pi i\omega' + 4\pi^2 D(\tilde{\epsilon}, z)(k_2')^2)\hat{T}(k',\omega') = \hat{T}_0(k')$$

where $D(\tilde{\epsilon}, z) = D_A(\tilde{\epsilon})$ from (3.29)A for $\tilde{\epsilon} > 0, z > 2$ and $D(\tilde{\epsilon}, z) = D_B(\tilde{\epsilon}, z)$ from (3.29)B for $\tilde{\epsilon} > 2 - z, z < 2$.

Thus, with the correspondence in (3.19)B, the eddy diffusivity equation predicted by R-N-G is

$$(3.32) \qquad \frac{\partial \bar{T}}{\partial t} = D(\tilde{\epsilon}, z)\bar{T}_{yy}$$
$$\bar{T}|_{t=0} = T_0(x)$$

where the formulas for $D(\tilde{\epsilon}, z)$ have already been given in (3.29) for $\tilde{\epsilon} > \max\{0, x - z\}$. The anomalous scaling exponents in (3.30) and the eddy diffusivity equation in (3.32) complete the predictions of the R-N-G from [26] on the simplified model.

SECTION 4: COMPARISON OF THE APPROXIMATE AND EXACT RENORMALIZATION THEORY FOR THE MODEL PROBLEM

We have established in Section 3 that the R-N-G theory of Yakhot and Orszag for the model problem in (2.15) predicts a "phase diagram" with three distinct regions of renormalization as depicted in Figure 2; this should be compared with the five regions of rigorous renormalization in the phase diagram from Figure 1.

The approximate R-N-G theory of Yakhot and Orszag exactly reproduces the rigorous theory of eddy diffusivity for the model problem in Region I — the regime of mean field theory. Both the diffusive scaling, $\rho(\delta) = \delta$ for Region I and the diffusion equation for $\bar{T}$ are reproduced exactly. As noted earlier in Section 3B), this is an expected result for the special model problem; furthermore, this fact indicates that the model problem is a very good test for this R-N-G method since the philosophy of this approximate R-N-G theory is to mimic Wilson's first order $\tilde{\epsilon}$-expansion from critical phenomena (see [18]) to make predictions in anomalous regimes from expansions based on the cross-over behavior at the boundary of mean field theory.

Region A is defined by the inequalities $\tilde{\epsilon} > 0$ and $z > 2$. As computed in (3.30) of Section 3C), in Region A, the approximate R-N-G theory of Yakhot and Orszag predicts the time rescaling functions,

$$(4.1)A \qquad \rho_A(\delta) = \delta^{1-\frac{\tilde{\epsilon}}{2}}$$

and the renormalized diffusivity equation for $\bar{T}$,

$$(4.1)B \qquad \bar{T}_t = D_A(\tilde{\epsilon})\bar{T}_{yy} .$$

with the coefficient $D_A(\tilde{\epsilon})$ given in (3.29)A. The *approximate renormalization theory for Region A coincides with the approximate renormalization theory* of Yakhot and Orszag applied to (2.15) *with the steady velocity field* in (2.16)A. Similar formulas as in (4.1) are valid at the boundary where $z = 2$ (see Section 3C)).

Region B is defined by the inequalities $\tilde{\epsilon} > 2 - z$ and $z < 2$. As calculated in Section 3C) in Region B, the approximate R-N-G theory of Yakhot and Orszag predicts the time rescaling function

$$(4.2)A \qquad \rho_B(\delta) = \delta^{\frac{4-\tilde{\epsilon}-z}{2}}$$

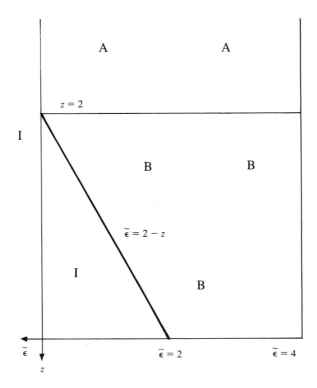

Figure 2: The phase diagram describing the three regions of renormalization in the $(\tilde{\epsilon}, z)$ upper half plane predicted by the Yakhot-Orszag renormalization theory applied to the model problem

and the renormalized diffusivity equation for $\bar{T}$,

$$(4.2)B \qquad\qquad \bar{T}_t = D_B(\tilde{\epsilon}, z)\bar{T}_{yy} \ ,$$

with the coefficient $D_B(\tilde{\epsilon}, z)$ given in $(3.29)B$.

   Next we comment on the exact renormalization theory versus the approximate renormalization theory in Regions A and B. Region A overlaps with Region V and Region III in the exact renormalization theory. In Region A, the approximate renormalization theory predicts the single scaling function in $(4.1)A$ while the rigorous theory gives the two different scaling functions $\rho(\delta) = \delta^{\frac{1}{1+\frac{\tilde{\epsilon}}{2}}}$ for $0 < \tilde{\epsilon} < 2$ for Region V and $\rho(\delta) = \delta^{1-\frac{\tilde{\epsilon}}{4}}$ for Region III. Region V is adjacent to the region of mean field theory and since

$$\frac{1}{1+\frac{\tilde{\epsilon}}{2}} = 1 - \frac{\tilde{\epsilon}}{2} + 0(\tilde{\epsilon}^2) \ ,$$

the predicted exponent of rescaling by the approximate R-N-G theory in Region A coincides with the exponent in (2.21) within $0(\tilde{\epsilon}^2)$ — this is the expected behavior of a theory based on Wilson's $\tilde{\epsilon}$-expansion. Of course there are large errors in the scaling exponents for $\tilde{\epsilon} > 2$. The renormalized diffusion equation in (4.1)B) predicted by the R-N-G theory of Yakhot and Orszag for Region A is a rather poor approximation to the rigorous renormalized diffusivity equations discussed earlier in (2.20)B) and (2.23) for Regions III and V respectively. Even though Region V is adjacent to the boundary of mean field theory, the rigorous renormalized diffusivity equation is nonlocal with broad band random diffusivity and the simple local diffusion equation predicted by the theory of Yakhot and Orszag is a poor approximation. In Region III, the Green's function for (4.1)B) is a poor approximation for the explicit Green's function for (2.20)B) which involves diffusion with a time-dependent coefficient. Incidentally, we remark that the simple local diffusion equation in (2.20)B) is not an artifact of the model, the rigorous renormalization theory in R^d for Region III necessarily has a similar structure ([7]).

In Region B, the approximate renormalization theory of Yakhot and Orszag yields much better results as a theory for renormalized eddy diffusivity. The Region B includes Region II, Region IV, and part of Region III from the rigorous renormalization theory (see Figures 2 and 3). In Region II, the predicted exponent in (4.2)A) coincides exactly with the rigorous exponent in (2.24)A); furthermore, the approximate renormalized diffusion equation in (4.2)B) has the same form as the rigorous renormalized diffusion equation in (2.24)B) — the coefficients of renormalized diffusivity $D_2(\tilde{\epsilon}, z)$ and $D_B(\tilde{\epsilon}, z)$ are also equal so the R-N-G method is exact in Region II!! For the velocity statistics in (3.6) the rigorous exact renormalized coefficient is readily computed and established to be equal to $D_B(\tilde{\epsilon}, z)$ from (3.29) by following the calculation on Page 408 of [1]; unfortunately, the constant for Region II reported in [1] is incorrect by a factor of four.

There is a special feature of the rigorous renormalization in Region II which provides an explanation for this spectacular success. The boundary of mean field theory in Region B is given by $\tilde{\epsilon} = 2 - z$ and Region II is adjacent to this boundary (see Figure 2); the rigorous scaling exponent in (2.24)A) is linear in $\tilde{\epsilon} - 2 + z$ so the first order $\tilde{\epsilon}$-expansion in the approximate R-N-G procedure should reproduce this exponent exactly; in fact the approximate theory achieves this in Region II with remarkable accuracy for the renormalized Green's function. However, the rigorous renormalization theory for Region IV in (2.25) and Region III in (2.20) show that the approximate R-N-G theory yields incorrect scaling exponents and a poor approximation for the renormalized Green's function in the parts of Region B excluding Region II in a fashion already discussed for Region A. It is amusing that the analogue of the Kolmogoroff spectrum in the model problem sits at the boundary between Region II and Region III with the values $\tilde{\epsilon} = \frac{8}{3}$ and $z = 2/3$ (see Remark 2 above) — this is a boundary point for good approximation by the $\tilde{\epsilon}$-expansion in Region B. Similar boundary phenomena are believed to occur for the Kolmogoroff spectrum in the approximate R-N-G theory for the Navier-Stokes equations (see [26]).

REFERENCES

[1] AVELLANEDA, M. AND MAJDA, A., *Mathematical models with exact renormalization for turbulent transport*, Comm. Math. Phys., 131 (1990), pp. 381–429.

[2] AVELLANEDA, M. AND MAJDA, A., *Approximate and Exact Renormalization Theories for a Model for Turbulent Transport*, (submitted to Phys. of Fluids, May 1991).

[3] AVELLANEDA, M. AND MAJDA, A., *Mathematical models with exact renormalization for turbulent transport II: non-Gaussian statistics, fractal interfaces, and the sweeping effect*, (submitted to Comm. Math. Phys., May 1991).

[4] AVELLANEDA, M. AND MAJDA, A., *An integral representation and bounds on the effective diffusivity in passive advection by laminar and turbulent flows*, accepted, Comm. Math. Phys., December 1990.

[5] AVELLANEDA, M. AND MAJDA, A., *Stieltjes integral representation and effective diffusivity bounds for turbulent transport*, Phys. Rev. Lett., 62 (1989), pp. 753–755.

[6] AVELLANEDA, M. AND MAJDA, A., *Homogenization and renormalization of multiple-scattering expansions for Green functions in turbulent transport*, Published in Composite Media and Homogenization Theory, Birkhauser, Boston, Edited by Dal Mao and Dell' Antonio, 1990, pp. 13–3.

[7] AVELLANEDA, M. AND MAJDA, A., *Super-Ballistic renormalization for turbulent transport in R^d*, in preparation.

[8] AVELLANEDA, M. AND MAJDA, A., *Renormalization Theory for Eddy Diffusivity in Turbulent Transport*, Phys. Rev. Letters, (submitted June 1990).

[9] BATCHELOR, G. K., *The Theory of Homogeneous Turbulence*, Cambridge Science Classics, Cambridge University Press, 1982.

[10] FORSTER, D., NELSON, D., AND STEPHEN, M., Phys. Rev. A., Vol. 16, #2,, pp. 732-749, 1977.

[11] KHAS'MINSKII, R.Z, *On the stability of nonlinear stochastic systems*, J. Appl. Math. Mech., 30 (1966), pp. 1082–1089.

[12] KHAS'MINSKII, R. Z., *Stochastic Stability of Differential Equations*, Sijthoff and Noordhoff, Alplen aan den Rijn, Netherlands, 1980.

[13] KOCH, D. AND BRADY, J., *Phys. Fluids A, 1.*

[14] KOCH, D. AND BRADY, J., J. Fluid Mech., 180 (1987), pp. 387–403.

[15] KRAICHNAN, R., *Complex Systems 1*, 1987, pp. 805–820.

[16] KRAICHNAN, R., Phys. Fluids, 6, #4 (1965), pp. 575–598.

[17] KUBO, R., J. Math. Phys., 4 (1963), pp. 174–178.

[18] MA, S. K., *Modern theory of Critical Phenomena*, Benjamin, Reading, Mass, 1976.

[19] MCCOMB, W. D., *The Physics of Fluid Turbulence*, Oxford Engineering Science Series #25, Clarendon Press, Oxford.

[20] MCLAUGHLIN, D., PAPANICOLAOU, G., AND PIRONNEAU, O., SIAM J. Appl. Math. Vol. 45 (1985), pp. 780–807.

[21] OELSCHLÄGER, K., Annals of Prob. 16, (3) (1988), pp. 1084–1126.

[22] PAPANICOLAOU, G. AND VARADHAN, S.R.S., *Boundary value problems with rapidly oscillating coefficients*, in Random Fields, Colloq. Math. Soc. Janos Bolyai, 27, J. Fritz, J. L. Lebowitz, and D. Szasz, editors, North Holland, Amsterdam, 1982, pp. 835–873.

[23] PAPANICOLAOU, G. AND KOHLER, W., *Asymptotic theory of mixing stochastic ordinary differential equations*, Comm. Pure Appl. Math., 27 (1974), pp. 311–350.

[24] ROBERTS, P.H., J. Fluid Mech., Vol. 11 (1961), pp. 257–273.

[25] ROSE, H. A., J. Fluid Mech., Vol. 81, #4 (1977), pp. 719–734.

[26] YAKHOT, V. AND ORSZAG, S., *Renormalization group analysis of turbulence. I basic theory*, J. Sci. Comp., 1 (1986), pp. 3–51.

[27] YAKHOT, V. AND ORSZAG, S., *Analysis of the ϵ-expansion in turbulence theory: approximate renormalization group for diffusion of a passive scalar in a random velocity field*, preprint (October 1988).

Appendix: The Finite Band Mode Elimination Scheme of Yakhot-Orszag

Our objective in this appendix is to establish the validity of the finite mode elimination algorithm in (3.15), (3.16) under the three assumptions from (3.13):

$(A.1)$
1) First Order High Wave Number Mode Elimination
2) Partial Averaging of High Wave Number Velocity Statistics
3) Principle of Distant Interaction.

Thus we need to establish the following *Equations* with *Enhanced Diffusivity after Finite Mode Elimination*:
$(A.2)$

Assume that $\hat{T}(k,\omega)$ involves only modes with

$|k_1| \leq \Lambda_0 e^{-r} = \Lambda(r)$ and satisfies

the integral equation,

$$(2\pi i\omega + 4\pi^2 \kappa_0 k_1^2 + 4\pi^2 \kappa k_2^2)\hat{T}(k,\omega)$$

$$= -2\pi i k_2 \int\int \hat{v}(k_1 - \tilde{k}, \omega - \tilde{\omega})\hat{T}(\tilde{k}, k_2, \omega)d\tilde{k}d\tilde{\omega} + \hat{T}_0^\delta$$

where $\hat{v}(k_1, \omega)$ vanishes for $|k_1| > \Lambda_0 e^{-r}$ and κ

is a general positive diffusion coefficient. If the modes k_1 for

$e^{-\tilde{r}}\Lambda(r) \leq |k_1| \leq \Lambda(r)$ are eliminated using the three approximations in (A-1),

then $\hat{T}^< = \chi^{r,\tilde{r}}\hat{T}$ satisfies the augmented viscosity equation

$(A.3)$
$$2\pi i\omega + 4\pi^2 \kappa_0 k_1^2 + 4\pi^2(\kappa + \Delta\kappa)k_2^2)\hat{T}^<$$

$$= -2\pi i k_2 \int\int \hat{v}^<(k_1 - \tilde{k}, \omega - \tilde{\omega})\hat{T}^<(\tilde{k}, k_2, \tilde{\omega})d\tilde{k}d\tilde{\omega} + \hat{T}_0^\delta$$

where $\hat{v}^< = \chi^{r,\tilde{r}}\hat{v}$ and

$(A.4)$
$$\Delta\kappa = 2 \int_{e^{-\tilde{r}}\Lambda(r)}^{\Lambda(r)} (2\pi a|k|^z + 4\pi^2 \kappa_0|k|^2)^{-1}|k|^{1-\tilde{\epsilon}}dk .$$

For simplicity in exposition, we have dropped all of the subscripts involving r in rewriting (3.15) and (3.16) in (A.2), (A.3), and (A.4) above.

We have already described in (3.10) – (3.12) the procedure involving the first order high wave number mode elimination in 1) of (A.1). For the specific integral equation in (A.2), the first order perturbation equation for $\hat{T}^>$ becomes

$(A.5)$
$$\hat{T}^>(k,\omega) =$$

$$g_0(k,\omega)(-2\pi i k_2 \int\int \hat{v}(k_1 - \tilde{k}, \omega - \tilde{\omega})\hat{T}^<(\tilde{k}, k_2, \tilde{\omega})d\tilde{k}d\tilde{\omega})$$

where the right hand side is evaluated for k_1 with $e^{-\tilde{r}}\Lambda \leq |k_1| \leq \Lambda$. Here $g_0(k,\omega)$ is the unperturbed Green's function for the integral equation in (A.2), i.e.

$(A.6)$
$$g_0(k,\omega) = (2\pi i\omega + 4\pi^2 \kappa_0 k_1^2 + 4\pi^2 \kappa k_2^2)^{-1} .$$

With the equation for $\hat{T}^>$ in (A.5) and the specific integral equation in (A.2), the equation in (3.12) becomes

$$g_0(k,\omega)^{-1}\hat{T}^<(k,\omega =$$

$$\left\{\hat{T}_0^\delta(k,\omega) - 2\pi i k_2 \int\int \hat{v}^<(k_1 - \tilde{k}, \omega - \tilde{\omega})\hat{T}^<(\tilde{k}, k_2, \tilde{\omega})d\tilde{k}d\tilde{\omega}\right\}$$

$$+ \left\{2\pi i k_2 \int\int \hat{v}^<(k_1 - \tilde{k}, \omega - \tilde{\omega})\hat{T}^>(\tilde{k}, k_2, \tilde{\omega})d\tilde{k}d\tilde{\omega}\right\}$$

(A.7)
$$+ \left\{- 2\pi i k_2 \int\int \hat{v}^>(k_1 - \tilde{k}, \omega - \tilde{\omega})\hat{T}^<(\tilde{k}, k_2, \tilde{\omega})d\tilde{k}d\tilde{\omega}\right\}$$

$$+ \left\{- 2\pi i k_2 \int\int \hat{v}^>(k_1 - \tilde{k}, \omega - \tilde{\omega})\hat{T}^>(\tilde{k}, k_2, \tilde{\omega})d\tilde{k}d\tilde{\omega}\right\}$$

$$\equiv \{1\} + \{2\} + \{3\} + \{4\}$$

where the right hand side in (A.7) is evaluated at wave numbers k_1 with $|k_1| \leq e^{-\tilde{r}}\Lambda$ and $\hat{T}^>(\tilde{k}, k_2, \tilde{\omega})$ is computed from the approximation in (A.5) so that

$$\{4\} =$$

(A.8)
$$- 4\pi^2 k_2^2 \int_{e^{-\tilde{r}}\Lambda \leq |\tilde{k}| \leq \Lambda} \int \hat{v}^>(k_1 - \tilde{k}, \omega - \tilde{\omega})g_0(\tilde{k}, \tilde{\omega})$$

$$\times \int\int \hat{v}(\tilde{k} - k', \tilde{\omega} - \omega')\hat{T}^<(k', k_2, \omega')dk'd\omega'd\tilde{k}d\tilde{\omega}$$

with a similar substitution of (A.5) into term $\{2\}$ which we will not need explicitly. The equations (A.7) and (A.8) are the explicit implementation for the integral equation in (A.2) of the mode elimination from 1) of (A.1). Next we assess the contributions of $\{2\}, \{3\}, \{4\}$ from (A.7) and (A.8) through the two other approximation procedures in (A.1) 2), 3).

First, we examine terms $\{2\}$ and $\{3\}$ which are easy to analyze and give us the opportunity to illustrate the *principle of distant interaction* from 3) of (A.1). In this principle (which is ubiquitous in the calculations from [26] and [27]), it is *assumed that whenever the external variables* k_1, k_2 and ω *are evaluated in the argument of a convolution with* $|k_1| < \Lambda e^{-\tilde{r}}$, *then if this is convenient the leading order approximation is given by* $k_1 \equiv 0, \omega \equiv 0$ (external variables in a convolution are variables that are not integrated). The intuition behind this assumption is that the ultimate interest for the iteration procedure is to apply it at large scales and long times, i.e. $|k_1| \to 0$ and $|\omega| \to 0$, so that perhaps, as a reasonable first approximation we have $k_1 \equiv 0$ and $\omega \equiv 0$; this is a crucial idea of Yakhot and Orszag from [26], [27] which facilitates explicit evaluation of eddy diffusivity (see for example, pages 10 and 11 of [26] or [27]). The present authors do not know a good mathematical justification for this principle in general and clearly this might be a significant source of error in the regions discussed in Section 4 where R-N-G has large errors when compared with the exact renormalization theory. Nevertheless, our goal here is to evaluate the R-N-G procedure from [26], [27] so we assume this principle here. Invoking the distant interact principle in term $\{2\}$ from (A.7), we obtain

$$(A.9) \qquad \{2\} \cong -2\pi i k_2 \int \int \hat{v}^<(-\tilde{k}, -\tilde{\omega}) \hat{T}^>(\tilde{k}, k_2, \tilde{\omega}) d\tilde{k} d\tilde{\omega} = 0$$

since $\hat{v}^<$ and $\hat{T}^>$ have disjoint supports. Similarly, for the term $\{3\}$ in (A.7) we can invoke the distant interaction principle to obtain by the same reasoning that $\{3\} \cong 0$. We remark that, alternatively, we can obtain that $\{3\} = 0$ by applying the statistical principle in 2) of (A.1) involving velocity averaging over modes with $|k_1| \leq \Lambda e^{-\tilde{r}}$ since $\langle v^> \rangle = 0$. Term $\{4\}$ yields the additional diffusivity in (A.3), (A.4) and we discuss this next. We decompose term $\{4\}$ given explicitly in (A.8) as follows:

$$\{4\} = \{4\}^< + \{4\}^>$$

with

$$(A.10) \qquad \begin{aligned} \{4\}^> = \\ -4\pi^2 k_2^2 \int_{e^{-\tilde{r}}\Lambda \leq |\tilde{k}| \leq \Lambda} \int \hat{v}^>(k_1 - \tilde{k}, \omega - \tilde{\omega}) g_0(\tilde{k}, \tilde{\omega}) \times \\ \int \int \hat{v}^>(\tilde{k} - k', \tilde{\omega} - \omega') \hat{T}^<(k', k_2, \omega') dk' d\omega' d\tilde{k} d\tilde{\omega} . \end{aligned}$$

We claim that $\{4\}^< = 0$; we observe that by definition $\{4\}^<$ involves the convolution of $\hat{v}^>$ with a function involving only $\hat{v}^<$ and $\hat{T}^<$ — thus, we invoke principle 2) of (A.1), take a statistical average over the high wave number modes, and utilize the fact that $\langle v^> \rangle = 0$ to obtain that $\{4\}^< = 0$. We compute $\{4\}^>$ explicitly by statistical averaging over the high wave number modes and utilizing the explicit form of the velocity statistics in (3.6). With a lengthy but straightforward calculation, we obtain

$$(A.10) \qquad \{4\}^> = -4\pi^2 k_2^2 \int_{\Lambda e^{-\tilde{r}} \leq |\tilde{k}| \leq \Lambda} |k_1 - \tilde{k}|^{1-\tilde{\epsilon}} a(\tilde{k}) d\tilde{k}$$

with $a(\tilde{k}, k_1, k_2, \omega)$ given by

$$(A.11) \qquad \begin{aligned} a(\tilde{k}, k_1 k_2, \omega) = \\ \pi^{-1} \int_{-\infty}^{\infty} a|k_1 - \tilde{k}|^z ((\omega - \tilde{\omega})^2 + a^2|k_1 - \tilde{k}|^{2z})^{-1} \times \\ (i(\omega - \tilde{\omega}) + 4\pi^2 \kappa_0(\tilde{k})^2 + 4\pi^2 \kappa k_2^2)^{-1} d\tilde{\omega} . \end{aligned}$$

Following Yakhot and Orszag (as in pages 9-10 [26] and also [27]), we invoke the distant interaction principle from 3) of (A.1) to evaluate the right hand side of (A.10) in the large scale limit of the external variables k_1, k_2, ω to obtain

$$(A.12) \qquad \{4\}^> \cong -4\pi^2 k_2^2 \int_{e^{-\tilde{r}}\Lambda \leq |k| \leq \Lambda} |k|^{1-\tilde{\epsilon}} a(k, 0, 0) dk .$$

From elementary explicit calculation (either by Plancherel's formula and Fourier analysis or residues) we compute that

$$(A.13) \qquad A(k,0,0,0) = (2\pi a|k|^z + 4\pi^2\kappa_0|k|^2)^{-1} .$$

Since we have shown that $\{2\} = 0, \{3\} = 0$, and the effect of $\{4\}$ is given in (A.12), (A.13), the reader can look back at equations (A.6), (A.7) and see that we have derived the equation in (A.3) with the enhanced diffusivity given by (A.4) through a systematic application of the three principles in (A.1). This completes our discussion in the appendix except for the following technical comment.

Remark:

For the anisotropic model problem in (2.15), the "sweeping effect" (see [27]) of the long wave number modes on the short wavelengths is negligible. To include the sweeping effect with the approximation in 1) of (A.1), the formula in (A.5) for $\hat{T}^>(k,\omega)$ is replaced by

$$(A.14) \qquad \begin{aligned} &\hat{T}^>(k,\omega) = \\ &G_0(k,\omega)(-2\pi i k_2 \int\int \hat{v}(k_1 - \tilde{k}, \omega - \tilde{\omega})\hat{T}^<(\tilde{k}, k_2, \tilde{\omega})d\tilde{k}d\tilde{\omega}) \end{aligned}$$

with G_0, the Green's function including the sweeping effect of large scales given by

$$(A.15) \qquad (G_0(k,\omega))^{-1} = (g_0(k,\omega))^{-1} + 2\pi i k_2 V^<$$

with $V^< = \int \hat{v}^<(k')dk'$. The important point for (A.15) is that the sweeping effect enters only as an *external* variable through (A.14) in term $\{4\}$. Thus, through the distant interaction principle, it is readily established that this effect is negligible — we leave the details to the interested reader.

WEAK AND STRONG TURBULENCE IN THE COMPLEX GINZBURG LANDAU EQUATION

J.D. GIBBON*

Abstract. We present analytical methods whereby weak and strong turbulence are predicted in the D dimensional complex Ginzburg Landau (CGL) equation

$$A_t = RA + (1 + i\nu)\Delta A - (1 + i\mu)A|A|^{2q}$$

on a periodic domain [0,1]. Strong (hard) turbulence is characterised by large fluctuations away from space & time averages while no such fluctuations occur in weak (soft) turbulence. In the $\mu - \nu$ plane, there are different areas where weak & strong behaviour can occur. In the strong case $(\mu, \nu \to \pm\infty, \mp\infty)$, the corresponding areas go out to the inviscid limit where the CGL equation becomes the NLS equation in which a finite time singularity occurs when $qD \geq 2$. A new infinite set of differential inequalities for the "lattice" of functionals

$$F_{n,m} = \int [|\nabla^{n-1}A|^{2m} + \alpha_{n,m}|A|^{2m[q(n-1)+1]}]d\underline{x}$$

enables us to construct large time upper bounds on the $F_{n,m}$. The occurrence of strong spiky turbulence is predicted for $qD = 2$ by showing that exponents of R in the upper bounds of $F_{n,m}$ & $\|A\|_\infty$ in the strong regions are dependent on the quantity $|\nu|$ which gets large in the inviscid limit. The critical value $qD = 2$ plays an important role: when $qD > 2$ the CGL equation has some similarities with the 3D Navier equations. A comparison is made between the two & the possibility of having a 1D system which mimics some limited features of the Navier Stokes equations is discussed.

§1. Introduction. One of the avowed aims of dynamical systems has been the solution of the problem of turbulence, although what this phenomenon has come to mean lies in the eye of the beholder. To most fluid dynamicists, the only real turbulence is the fine scale 3-dimensional turbulence which occurs at high Reynolds numbers, with an energy cascade and an inertial subrange. The number of degrees of freedom in 3D strong turbulence is clearly many orders of magnitude greater than in such phenomena as convection in a box (see references in [1]) where perhaps only a few spatial modes govern the dynamics. Both come under the heading of 'turbulence' but attempts to equate the two types of phenomena have led to much confusion.

Only in 2 spatial dimensions are the incompressible Navier Stokes equations understood analytically in the sense that there is a rigorous proof of the existence of a finite dimensional global attractor [2, 3, 4]. If G is the Grashof number, then the dimension of this attractor is bounded above by $cG^{2/3}(1+\log G)^{1/3}$. Computational methods are generally good enough to resolve the smallest scale in a 2D flow (see refs in [5]) and, for 2D homogeneous decaying turbulence, the vorticity obeys a maximum principle. No such maximum principle is known to exist in 3D and regularity remains to be proved in this case. Furthermore, numerical resolution of the smallest scale in a fully turbulent 3D flow is still a long way off.

*Department of Mathematics, Imperial College, London SW7 2BZ, UK

There is yet one more important phenomenon which, as yet, lies in the realm of the unknown & that is the question of finite time singularities in the Euler equations. These are the inviscid limit of the Navier Stokes equations and it is generally believed that, starting from smooth initial data, the 2D Euler equations do not blow up in finite time while the 3D equations do. However, there is no proof of either of these conjectures (see [6]). Studies in Zakharov's equations have done much to highlight the role played by finite time singularities in plasma turbulence [7,8] but, unfortunately, these equations have yielded little to analytical endeavour and so these studies are of little help in trying to understand the Euler equations.

In order to attempt to get a better grip on the tantalising phenomena displayed by the Navier Stokes equations, it is a useful exercise to see whether it is possible to mimic some limited features of the 3D Navier Stokes equations with a different PDE system which displays similar functional properties but in a lower spatial dimension. This exercise, however, must obviously be limited by the fact that simpler models in lower dimensions cannot display the vortex stretching properties displayed by the 3D Navier Stokes equations, although the lowering of the spatial dimension does make it easier to compute the dynamics.

One equation which we will show has some of the desired properties is a version of the complex Ginzburg Landau equation (CGL) in in D spatial dimensions

$$(1.1) \qquad A_t = RA + (1 + i\nu)\Delta A - (1 + i\mu)A|A|^{2q}$$

We use periodic boundary conditions on the domain $[0, 1]$. R, μ and ν are real parameters with $R \geq 0$ and μ and ν of either sign. It is *not* our intention here to treat it in its physical context: Indeed we shall divorce it from that entirely and treat it purely as an abstract model. In this sense, we do not pretend that it describes true fluid turbulence. Our intention in using it is to try and mimic limited features of the Navier Stokes equations with an equation over which we have more analytical control. Specifically we would like to have the following properties:

(i) a finite time singularity in the inviscid limit

(ii) an area of the parameter space, going out to the inviscid limit, where strong turbulence occurs.

(iii) a transition from strong turbulence to a much weaker form of turbulence as dissipation increases.

(iv) the possibility that these phenomena may occur in 2D or even 1D thereby making numerical studies much easier.

(v) the occurrence of an inertial subrange in the strong region.

What we have loosely called 'strong' and 'weak' turbulent behaviour needs some explanation. We define strong or hard turbulence to be a phenomenon where large fluctuations away from temporal & spatial averages occur. If this occurs, then solutions must be spatially & temporally narrow (i.e. spiky) with a consequent cascade of energy to high k. The spiking of the solution would rationally have its source in the finite time singularity in the inviscid limit trying to force the solution to blow up, only for the dissipation to finally control the system and bring it back down again. In contrast, weak or soft turbulence maybe chaotic but would display no great fluctuations away from spatial & temporal averages. In such a system, one naturally expects strong turbulent behaviour to occur near the inviscid limit where dissipation is small and the weak turbulent behaviour where dissipation is stronger.

We note straight away that condition (i) is fulfilled in the case of the CGL equation. The inviscid limit ($R \to 0$, $|\nu|$, $|\mu| \to \infty$) produces the NLS equation in D dimensions

$$(1.2) \qquad iA_t = \nu \Delta A - \mu A |A|^{2q}$$

which is well known to blow up in finite time [9] provided $qD \geq 2$ and provided the energy,

$$(1.3) \qquad E = \int \left[|\nabla A|^2 + \frac{\mu}{2\nu} |A|^{2(q+1)} \right]$$

is negative. Since E is a constant of the motion, this can be set negative by suitably chosen initial data, provided μ and ν are of opposite sign. It is particularly the so-called "critical case" $Dq = 2$ which interests us here because it allows us to look at the $D = 2$ (with $q = 1$) and $D = 1$ (with $q = 2$) cases together. When it is super-critical, for instance when $qD = 3$, then the $D = 1$, $q = 3$ case behaves in much the same way as the $D = 3$, $q = 1$ case. Consequently, we have a 1D system which may, to some degree, mimic the behaviour of a 3D system, although it is also the criticality or supercriticality of the system which is as important as the dimension itself.

This paper is a summary of the results of the papers by Bartuccelli, Constantin, Doering, Gibbon & Gisselfält [10] and Doering, Gibbon and Levermore [11]. Only sketches of theorem proofs will be given here as they can be found in full in these two references.

§2. A lattice theorem for the CGL equation. The CGL equation, written in the form given in (1.1) has a natural symmetry scaling: if A scales like λ then ∇ scales like λ^q. This motivates us to consider functionals whose terms are of equal weight [10,11]:

$$(2.1) \qquad F_{n,m} = \int [|\nabla^{n-1} A|^{2m} + \alpha_{nm} |A|^{2mn(q)}] \, d\underline{x}$$

where $\alpha_{nm} > 0$ for all m, $n \geq 1$ with $n(q) = q(n - 1) + 1$. The $F_{n,m}$ can be thought of as defining a 'lattice' where one steps up in n and along in m and is a generalization of the ladder idea described in [10]. The ∇ operator is used in its general sense. Now we define the quantity

$$(2.2) \qquad \beta_{m,\nu} = m - |1 + i\nu|(m - 1)$$

which appears in the following theorem [11]:

THEOREM 1. *For $m, n \geq 1$ and provided m and ν are chosen such that $\beta_{m,\nu} > 0$, then*

$$(2.3a) \qquad \dot{F}_{n,m} \leq [2mn(q)R + c_{nm}\|A\|_\infty^{2q}]F_{n,m} - b_{nm}F^{1+1/m}/F_{n-1,m}^{1/m}$$

where $c_{1,1} = 0$ and

$$(2.3b) \qquad b_{n,m} = m \min \left\{ \frac{\beta_{m,\nu}}{(2m-1)^2} \; ; \; \left(\frac{\alpha_{n-1,m}}{\alpha_{n,m}} \right)^{1/m} \right\}$$

Proof. This is a generalisation of the ladder theorem proved in [10] and is given in [11]. We will give only a short sketch of some of the salient points. Let us define

$$(2.4) \qquad G_{n,m} = \int |A|^{2mn(q)} \qquad J_{n,m} = \int |\nabla^{n-1}A|^{2m}$$

Following [10], it is easily seen that

$$(2.5) \qquad \frac{1}{2m} \dot{J}_{n,m} \leq RJ_{n,m} + c\|A\|_\infty^{2q}J_{n,m} - \beta_{m,\nu} \int |\nabla^n A|^2 |\nabla^{n-1}A|^{2(m-1)}$$

and

$$(2.6) \qquad \begin{aligned} \frac{1}{2mn(q)} \dot{G}_{n,m} &\leq RG_{n,m} - \int |A|^{2[mn(q)+q]} - \\ &[n(q)m - |1 + i\nu|(n(q)m - 1)] \int |\nabla A|^2 |A|^{2[n(q)m-1]} \end{aligned}$$

To deal with the negative definite term in (2.5) we first prove

LEMMA 1. *For $m, n \geq 1$*

$$(2.7) \qquad -\int |\nabla^n A|^2 |\nabla^{n-1}A|^{2(m-1)} \leq -\frac{1}{(2m-1)^2} J_{n,m}^{1+1/m}/J_{n-1,m}^{1/m}$$

Proof. After an integration by parts ($m \geq 1$) on $J_{n,m}$ we find

$$(2.8) \qquad J_{n,m}^2 \leq (2m-1)^2 \left[\int |\nabla^n A| \, |\nabla^{n-1}A|^{2(m-1)}|\nabla^{n-2}A| \right]^2$$

and so

$$-\int |\nabla^n A|^2 |\nabla^{n-1}A|^{2(m-1)} \leq -\frac{1}{(2m-1)^2} J_{n,m}^2 \left[\int |\nabla^{n-1}A|^{2(m-1)}|\nabla^{n-2}A|^2 \right]^{-1}$$

Using a Hölder inequality on the denominator we have the required result. □

To deal with the equivalent term in (2.6) we prove

LEMMA 2. For $n(q), m \geq 1$

$$(2.9) \qquad -\int |A|^{2(mn(q)+q)} \leq -G_{n,m}^{1+1/m}/G_{n-1,m}^{1/m}$$

Proof.

$$(2.10) \qquad G_{n,m} = \int |A|^{2mn(q)} \equiv \int |A|^P |A|^Q$$

where $P + Q = 2mn(q)$. Now we use a Hölder inequality on (2.10) to get

$$(2.11) \qquad G_{n,m} \leq \left[\int |A|^{P(m+1)/m} \right]^{m/(m+1)} \left[\int |A|^{Q(m+1)} \right]^{1/(m+1)}$$

and then choose P and Q such that

$$(2.12) \qquad P(m+1) = 2m(mn(q)+q) \qquad Q(m+1) = 2m(n(q)-q)$$

From (2.11), we have the result. To be sure $Q \geq 0$ we need $q \geq 1/2$. $\square$

The coefficient $\beta_{n(q)m,\nu}$ in (2.6) is not necessarily positive when $\beta_{m,\nu} > 0$ & use of an interpolation inequality, as in [10], shows that this term is bounded above by $\|A\|_\infty^{2q} F_{n,m}$. Taking the combination of the inequalities (2.5) and (2.6) to form $F_{n,m}$ and using the Schwarz inequality to piece together the results of Lemmas 1 and 2, the theorem is proved. $\square$

We now proceed to deal with the $\|A\|_\infty^{2q}$ term in Theorem 1.

THEOREM 2.

$$(2.13) \qquad \|A\|_\infty^2 \leq c \left\{ F_{n,m}^{\frac{2}{2n(q)m-Dq}} + F_{n,m}^{\frac{1}{n(q)m}} \right\}$$

provided $n > 1 + D/2m$.

Proof. Using the interpolation inequality

$$(2.14) \qquad \|A\|_\infty \leq c \|\nabla^{n-1} A\|_{2m}^a \|A\|_{2mn(q)}^{1-a} + \|A\|_2$$

so

$$(2.15) \qquad \|A\|_\infty^2 \leq c \, J_{n,m}^{a/m} G_{n,m}^{(1-a)/mn(q)} + \|A\|_2^2$$

where $\qquad a = D[2n(q)m(n-1) - D(n(q)-1)]^{-1}$

the result follows from the use of (2.17). The restriction $a < 1$ gives $n > 1 + D/2m$.

THEOREM 3.

(2.16a)
$$\varlimsup_{t \to \infty} F_{n,m} \le c \left(\varlimsup_{t \to \infty} F_{n-1,m} \right)^{\frac{[2mn(q)-Dq]}{[2mn(q)-Dq-2mq]}}$$

provided

(2.16b)
$$n > 1 + D/2m + (q-1)/q.$$

Proof. The result follows immediately from Theorem 1. The condition (2.16b) arises from the necessity of keeping the denominator in the exponent of (2.16a) positive. The dominant terms for large R only have been included. Before we discuss the minimum value of n which satisfies (2.16b) for given values of D and q, we need to state one more theorem:

THEOREM 4.

(2.17a)
$$\varlimsup_{t \to \infty} \int |A|^{2(1+\eta)} \le R^{(1+\eta)/q} \qquad \eta \ge 0$$

where

(2.17b)
$$|1+i\nu| \le 1 + 1/\eta$$

Proof.

From (2.6) we have

(2.18a)
$$\frac{1}{2mn(q)} \dot{G}_{n,m} \le R G_{n,m} - G_{n,m}^{1+1/m}/G_{n-1,m}^{1/m}$$

provided $|1+i\nu| \le n(q)m/[n(q)m-1]$. Hence there exists a ball

(2.19)
$$\varlimsup_{t \to \infty} G_{n,m} \le R^{n-1+1/q}$$

where $m = 1$ and n is chosen such that $n(q) = 1 + \eta$. Hence $n(q) - 1 = q(n-1) = \eta$. To satisfy this but also to ensure that $\beta_{m,\nu} > 0$ in Theorem 1, one convenient choice of η is

(2.20)
$$\eta = 1/|\nu|$$

which, in (2.19), finishes the proof. ☐

We now return to (2.16b) which places a condition on the "starting point" of the lattice. Rewriting it as

(2.21)
$$n > 2 + (Dq - 2m)/2mq$$

we can see that we are able to choose $n = 2$ provided

(2.22)
$$Dq < 2m$$

With $n = 2$ the starting point of the lattice $(F_{n-1,m})$ is

(2.23)
$$F_{1,m} = \int |A|^{2m}$$

We illustrate how (2.17) and (2.21) help with 3 special cases:

(a) $Dq = 1$ Choose $m = 1$ in (2.21) & the starting point is $F_{1,1}$. This is the L^2 norm which is bounded above for large times. Hence we have control over the whole ladder for $m = 1$. This case includes the standard 1D q=1 CGL equation.

(b) $Dq = 1$ Take $m = 1 + \eta$ $(\eta > 0)$ in (2.21) & the starting point is $F_{1,1+\eta}$ which is given by Theorem 4 with the choice of η depending on the value of ν.

(c) $Dq = 3$ The use of Theorem 4 is valid provided $m > 3/2$ (or $\eta > 1/2$). Equation (2.21) means that this is only valid in the region $|\nu| < \sqrt{8}$.

Case (c) simultaneously includes the $D = 3$, $q = 1$ case & the $D = 1$ $q = 3$ case. We therefore have control over the rungs of the ladder only in the region $|\nu| < \sqrt{8}$ for all μ. Outside of this region we have only the result of Theorem 4 which is not sufficient. In this region, control over the H^1-norm is necessary which we do not have a priori.

§3 **Estimates and time averages in the $\mu - \nu$ plane.** We shall begin this section by considering the quantity

(3.1)
$$F_{2,1} = \int [|\nabla A|^2 + \alpha_{21}|A|^{2(q+1)}]$$

We can find easily an upper bound on its time average by taking the CGL equation, multiplying by A^*, taking the real part, integrating over space and then taking the time average. We get

(3.2a)
$$\langle F_{2,1} \rangle \leq cR\langle \|A\|_2^2 \rangle \leq c\, R^{1+1/q}$$

where the time average $\langle . \rangle$ is defined as

(3.2b)
$$\langle f(t) \rangle \leq \lim_{t \to \infty} \limsup_{f(0)} \frac{1}{t} \int_0^t f(s)ds$$

We note that the upper bound on $\langle F_{2,1} \rangle$ in (3.2a) is uniform in μ and ν. For the critical case $qD = 2$, the bottom point of the lattice is $F_{2,1}$ so we need to compute $\overline{\lim}_{t \to \infty} F_{2,1}$ in different areas of the $\mu - \nu$ and then compare with (3.2a). To achieve this, it turns out that we can conveniently divide the $\mu - \nu$ plane into two specific areas.

§**3.1 Weak turbulent behaviour.** As we saw at the end of the last section, when $Dq = 2$, it is possible to find an upper bound on $\varlimsup_{t\to\infty} F_{2,1}$ directly from (2.16a,b) and Theorem 4. Since η is ν-dependent, these bounds will have ν in the exponent of R. For weak turbulence we need to to better than this. To show that the upper bound on $\varlimsup_{t\to\infty} F_{2,1}$ can be close to that on $\langle F_{2,1}\rangle$ in certain areas of the $\mu - \nu$ plane, it is necessary to be much more precise with the quantity $F_{2,1}$ and not pull out the amplitude in L^∞, as in the lattice theorem. A careful and technical procedure, explained in [10], allows us to discard certain combinations of the nonlinear and the Laplacian terms which turn out to be negative definite. This, however, can only be done in a certain area of the $\mu - \nu$ plane (away from the inviscid limit). We restrict ourselves to the case $q = 1$ for simplicity. The following theorem is stated and proved in [10]:

THEOREM 5. *If μ and ν lie in the unshaded region of the $\mu - \nu$ plane (see Figures 1 and 2); that is, if*

$$(3.3a) \qquad \nu > -\frac{\mu\sqrt{3}}{\mu - \sqrt{3}} \qquad \mu > \sqrt{3}$$

$$(3.3b) \qquad \nu < -\frac{\mu\sqrt{3}}{\mu + \sqrt{3}} \qquad \mu < -\sqrt{3}$$

or

$$(3.3c) \qquad |\mu| < \sqrt{3}$$

then coefficients $\alpha_{2,1} > 0$ and $b_2 > 0$ exist such that for every D

$$(3.3d) \qquad \dot{F}_{2,1} \le 4R\, F_{2,1} - b_2 F_{2,1}^2/R \qquad \square$$

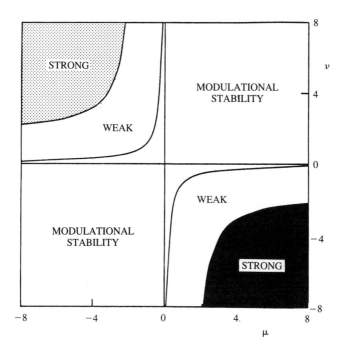

Figure 1: This summarises the behaviour of the $q = 1$, $D = 2$ CGL equation in the $\mu - \nu$ plane. The label "modulational stability" inside the inner hyperbolae refers to the Lange and Newell criterion [13] $\varepsilon = 1 + \mu\nu$ where the spatially homogeneous solution is stable (unstable) when $\varepsilon > 0 \ (< 0)$. This is discussed in detail in [10, 17]. The hyperbolae dividing the unshaded (weak) from the shaded (strong) regions are those given in Theorem 5 [12]. For $D = 1$ & $q = 2$, the results are essentially the same except the hyperbolae are shifted slightly. Estimates in the 'weak' region predict only weak departures from space & time averages. In the 'strong' region, large deviations away from averages are possible, indicating spatially & temporally localised intermittent spiking.

42

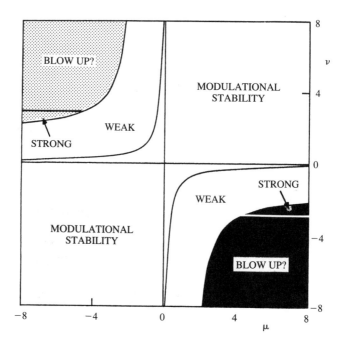

Figure 2: Behaviour of the CGL equation for $D = 3$
and $q = 1$. The designation of weak and strong areas is
the same as in Figure 1 except that the bounds in the
strong region are valid only for $\nu < |8|$ which goes out
into the shaded region as far as the anvil shaped areas
marked in the figure.

We conclude from (3.3) that in the unshaded region there is a ball

$$(3.4) \qquad \varlimsup_{t \to \infty} F_{2,1} \le c\, R^2 \qquad \varlimsup_{t \to \infty} \|A\|_\infty^2 \le c\, R^{4/(4-D)}$$

All the other $F_{n,1}$ can be computed as in [10]. The large time upper bound on $F_{2,1}$
compares favourably with the R^2 upper bound on $\langle F_{2,1}\rangle$ as also does the pointwise
bound of A compared to the L^2 bound. Furthermore, the exponents in both cases
are not dependent on ν or μ. Clearly, therefore, no massive excursions away from
averages can occur and spiky turbulence cannot happen. We would expect, at
most, 'weak' turbulence with possible chaotic states but with no cascade of energy
or inertial range.

§**3.2 Strong turbulent behaviour.** Since Theorem 5 is not valid in the
shaded region of the $\mu - \nu$ plane we need to find $\varlimsup_{t \to \infty} F_{2,1}$ for the rest of this
area. The use of (2.16a) & (2.16b) together with Theorem 4 give upper bounds

which are ν dependent[1]

(3.5a) $$\varlimsup_{t\to\infty} F_{2,1} \leq c\, R^{1/\eta+(q+1)/q} = c\, R^{|\nu|+(q+1)/q}$$

Furthermore, when $D = 2$ and $q = 1$, for example, then the large time pointwise bound on A becomes

(3.5b) $$\varlimsup_{t\to\infty} \|A\|_\infty^2 \leq c\, R^{1+1/\eta} = c\, R^{|\nu|+1}$$

Comparing the $\varlimsup\limits_{t\to\infty} F_{2,1}$ with its time average we see that the former can be *many* orders of magnitude greater than the latter if $|\nu| \gg 1$. This is precisely the behaviour for which we are looking in the inviscid limit. A comparison of $\|A\|_\infty$ with $\|A\|_2$ produces the same conclusion.

These two comparisons show that for solutions which achieve, or nearly achieve, the suprema, then the large excursions away from temporal & spatial averages which can occur must be in the form of temporally and spatially narrow spikes. An obvious physical explanation for this is that the finite time singularity in the inviscid limit tries to make the solution blow up. The amplitude and gradients increase and the solution grows and narrows into a spike but finally the dissipative terms catch up with it and bring it back down. The spike must be temporally and spatially narrow to avoid violating the bounds on the averages.

§4 Conclusion.

§4.1 An approximate lower bound on F_2 for a short time interval.
There is a mechanism through which we can obtain approximate lower bounds on some of the quantities involved. It is well known that in the NLS equation, there is a critical value of the $\|A\|_2^2$ which it is necessary for a solution to achieve before which a finite time singularity can appear [7,8]. Assuming that the energy E is negative, with μ & ν of opposite sign, we have

(4.1a) $$\|\nabla A\|_2^2 \leq |\mu/2\nu|\, \|A\|_{2(q+1)}^{2(q+1)}$$

(4.1b) $$\leq c|\mu/2\nu|\, \|\nabla A\|_2^{qD}\|A\|_2^{2(q+1)-qD}$$

where we have used an interpolation inequality on the RHS of (4.1a). Hence, when $qD = 2$, there is a lower bound on $\|A\|_2^2 \geq M_0$. M_0 can be thought of as the 'critical mass' necessary to make E negative thereby producing a condition for a spike. While (4.1b) is true only for the NLS equation and not CGL, nevertheless we argue that if we are sufficiently close to the NLS equation (out in the shaded region) then we use M_0 as an approximate lower bound on $\|A\|_2^2$ for the CGL equation for short time intervals once the equation has built up enough energy to pass criticality in the L^2 norm. Since $\|A\|_2^2$ is bounded below then so are the higher L^p norms and so we can easily deduce, via an interpolation inequality, that

(4.2) $$M_0^{1/2} \leq \|A\|_p \leq c\|\nabla A\|_2^a\|A\|_2^{1-a}$$

---

[1] These can be slightly improved by a more precise handling of the nonlinear term [10].

where $2pa = D(p-2)$. Hence we need $2 < p < 2D/(D-2)$. Since $\|A\|_2$ is bounded both above and below and $\|A\|_p$ bounded below, we have a lower bound on $\|\nabla A\|_2$. Consequently, we can construct an approximate lower bound on F_2 valid only for the time interval where we have an approximate lower bound on $\|A\|_2$. This approximate procedure helps to illustrate how the critical mass needed to create the NLS singularity also gives information about the spike which forms in the data for CGL.

§4.2 **Remarks on minimum length scales.** The nature of this paper is such that it is not appropriate to give a detailed numerical study of the CGL equation. This has been done in some detail in 1D [14, 15], in less detail in 2D [16] and not at all in 3D. It is clear that the spiky region requires careful numerical handling because the length scales are much shorter. Indeed, the lattice gives some lower bounds on the minimum scale in the flow. We can construct a set of typical length scales

$$(4.3) \qquad \lambda_{n,m}^{-2m} \equiv \langle F_{n,m}/F_{n-1,m}\rangle$$

From Theorem 1 we can find an upper bound on this by dividing the lattice inequality through by $F_{n,m}$ and using Hölder's inequality. We can see that

$$(4.4) \qquad \lambda_{n,m}^{-2} \leq c_{n,m}\langle\|A\|_\infty^2\rangle + 2mn(q)R$$

Clearly, in the weak or soft regime (unshaded region of Figures 1 & 2), $c_{n,m} \equiv 0$ and so we have the minimum scale $\approx R^{-1/2}$. In contrast, in the strong or hard region we must keep the $\langle\|A\|_\infty^2\rangle$ term. The time average can be replaced by a limsup in the limit of large t but we then find that the exponent is $|\nu|$-dependent. Hence the minimum scale in terms of $\lambda_{n,m}$ is *many* orders of magnitude smaller in the strong or hard region than in the weak or soft region. We would expect this as most of the energy packs into very short scales on the order of the width of the narrowest spike.

Conventionally, the minimum scale of a system of volume L^D in D spatial dimensions is taken to be the dissipation length ℓ_{diss} which is related to the number of degrees of freedom N_{df} by $N_{\text{df}} \approx (L/\ell_{\text{diss}})^D$. One would expect that ℓ_{diss} would be approximately the same as the natural scale $\lambda_{n,m}$ with N_{df} identified as the attractor dimension [17–19]. This appears to be the case in the weak region as (see [10]) N_{df} is such that $\ell_{\text{diss}} \geq cR^{-1/2}$. However, in the strong region this is clearly not the case as $\ell_{\text{diss}} \gg \lambda_{n,m}$. This discrepancy arises because the time average procedure in the calculation of the attractor dimension averages out the short scale motion. The idea that ℓ_{diss} represents some smallest scale in the flow is therefore not true in the spiky region.

In terms of properties (i) to (v), we seem to have achieved (i), (ii) and (iii). Property (iii), which is a requirement that there exists a transition between the weak and strong regimes is what is observed in gaseous helium convection experiments [20]. Property (iv) is also fulfilled in the sense that the critical case $qD = 2$ contains the $D = 1$ $q = 2$ and the $D = 2$ and $q = 1$ cases simultaneously. Computing a

1D fifth order parabolic PDE is much easier than a 2D cubic one. In the weak region and within the area $|\nu| < \sqrt{8}$ the $D = 1$ $q = 3$ and $D = 3$ $q = 1$ cases are equivalent. We have control in L^p only up to L^3 and this allows us to prove regularity for $|\nu| < \sqrt{8}$ but not beyond. Outside this region when $|\nu| \geq \sqrt{8}$, we need control over the H^1-norm which is not yet available.

Property (v) is the final condition listed in §1. It turns out that preliminary work [21] indicates that an inertial range does exist for the $D = 1$ $q = 2$ case with R taken up to 1000 and $\nu = 30$, $\mu = -30$.

§**4.3 A comparison with the 3D Navier stokes equations.** The ladder/lattice structure elaborated in Theorem 1 also occurs for the incompressible Navier Stokes equations which we take on periodic boundary conditions on the domain [0,1] in the standard velocity notation with unit density, with ν as viscosity & a zero momentum condition $\int_\Omega \underline{u} \, d\underline{x} = 0$. We also take, for simplicity a C^∞ forcing function f.

$$(4.5) \qquad \underline{u}_t + (\underline{u} \cdot \nabla)\underline{u} = \nu\Delta\underline{u} - \nabla p + \underline{f} \qquad \text{div } \underline{u} = 0$$

Now we define

$$(4.6) \qquad H_n = \sum_{i=1}^{3} \int (\nabla^n u_i)^2 d\underline{x}$$

where the gradient operator ∇ takes its generalised definition. The following theorem has been proved in [22]:

THEOREM 6. *For $n \geq 1$ and for $D = 2\&3$, the H_n satisfy the two alternative differential inequalities*

$$(4.7a) \qquad \frac{1}{2}\,\dot{H}_n \leq -\frac{\nu}{2D}\frac{H_n^2}{H_{n-1}} + \nu^{-1}c_{n,1}H_n\|\underline{u}\|_\infty^2 + H_n^{1/2}\sum_i \|\nabla^n f_i\|_2$$

$$(4.7b) \qquad \frac{1}{2}\,\dot{H}_n \leq -\frac{\nu}{D}\frac{H_n^2}{H_{n-1}} + c_{n,2}H_n\|\nabla u\|_\infty + H_n^{1/2}\sum_i \|\nabla^n f_i\|_2$$

The consequences of Theorem 6 will be mentioned only in brief as here is not the place to repeat all the definitions and calculations on attractors for the 2D and 3D Navier Stokes equations which be found in [3, 4, 23]. However, as a brief summary it is not difficult to describe what results the ladder method can reproduce.

Firstly, we consider (4.7a) and use an interpolation inequality

$$(4.8) \qquad \|u\|_\infty^2 \leq c\, H_n^{D/2n}\|u\|_2^{(2n-D)/n} \qquad n \geq 2$$

where we know solutions are bounded in L^2 independently of the pressure [3, 4, 23]. Since we are restricted by $n \geq 2$ then $D < 2n$ for $D = 2,3$ so we have absorbing balls for all H_n *provided* one has control over H_1 for large times. The quantity H_1 is known to be bounded above by its initial value only when $D = 2$, since it is the vorticity, for which there is a maximum principle. No such upper bound on H_1 is known when $D = 3$ and one has to resort to assumptions such as constructing an 'internal' Reynolds number from $\|u\|_\infty$ [3]. Hence we have C^∞ attractor for $D = 2$ but nothing more than an L^2 bound when $D = 3$. The method, as it stands, generalises into differential inequalities the bootstrapping from one Sobolev space to the next which has been commonly employed over the years [23].

The consequences of Theorem 6 also bear on the work of Henshaw, Kreiss and Reyna [24]. If the forcing is switched off, then the natural set of length scales which arise from Theorem 6 [equation (4.7b)] are

$$(4.9) \qquad \ell_n \geq c_n \left[\frac{\nu}{\sup\limits_t \|\nabla u\|_\infty} \right]^{1/2} \equiv c_n \lambda_{\min}$$

As in [22], by integrating across the ladder and then looking at the Fourier coefficients of the $H_n \equiv \sum h(k) \exp(ikx)$, it can be easily seen that wavenumbers k larger than $\lambda_{\min}^{-1}$ decay (provided one assumes that H_1 is bounded). In this cas,e the decay rate is algebraic & not exponential, as in [24], because of the n dependence of the c_n. The authors in [24] show that in the limit of vanishing viscosity ν, Fourier coefficients decay *exponentially* fast for wave numbers larger than $\lambda_{\min}^{-1}$ provided $\sup\limits_t \|\nabla u\|_\infty$ is assumed to be bounded. Foias and Temam [25] have shown an exponential decay of Fourier coefficients by showing that solutions are analytic in time in a spatial Gevrey class of functions provided H_1 is bounded. This is always true in 2D but must be assumed in 3D.

The argument in [24], which is based on the assumption of the boundedness of $\sup\limits_t \|\nabla u\|_\infty$, can be turned into an argument identifying $\lambda_{\min}$ with the Kolmogorov length scale [24, 22]. The Kolmogorov length scale λ_{KG} is conventionally defined in terms of the energy dissipation rate ε and the viscosity ν:

$$(4.10) \qquad \lambda_{KG} \approx \nu^{3/4} / \varepsilon^{1/4}$$

as it is the only combination of ν and ε which has dimensions of length. The energy dissipation rate is $\varepsilon = 2\nu \sup\limits_t H_1$, so we then introduce the approximation that for weakly turbulent flows

$$(4.11) \qquad \sup\limits_t \|\nabla u\|_\infty \approx \sup\limits_t H_1^{1/2}$$

whence from (4.9) we find

$$(4.12) \qquad \lambda_{\min} \approx [\nu^{3/2} / \varepsilon^{1/2}]^{1/2} = \lambda_{KG}$$

We can summarise by saying that the condition needed to prove regularity for supercritical CGL when $|\nu|$ is large, i.e. control over the H^1 norm, is the same as that required for the 3D Navier Stokes equations. In the latter case this can be reduced to L^3 [3] but not in the former case. Hence, the necessary condition for regularity for the supercritical CGL equation in the limit of large $|\nu|$ is stronger than that for the Navier Stokes equations. By super-critical we mean $Dq > 2$ which obviously includes both the $D = 1$, $q = 3$ and $D = 3$, $q = 1$ cases. We have not, however, been able to find weak and strong areas in the Navier Stokes equations in the same way as the CGL equation as there are less parameters.

Acknowledgements. Most of this work is based on references [10, 11, 22, 21] & I am therefore indebted to my friends Michele Bartuccelli, Petre Constantin, Charlie Doering, Darryl Holm, Mac Hyman, Gregor Kovacic & David Levermore with whom it has been great fun to work the last few years.

REFERENCES

[1] LIBCHABER, A., Proc. Royal Soc. London 413A, 63, 1987 and *From chaos to turbulence in Benard convection*, published in Dynamical Chaos, Princeton University Press (1887) p. 63.

[2] CONSTANTIN, P., FOIAS, C. & TEMAM, R., Physica 30D, 284 (1988).

[3] CONSTANTIN, P., FOIAS, C., *The Navier Stokes equations*, Chicago University Press (1989).

[4] TEMAM, R., *Infinite dimensional dynamical systems in mechanics & physics*, Springer Applied Math Series 68 (1988).

[5] WEISS, N.O., Proc. Royal Soc. London 413A, 71, (1987) and *From chaos to turbulence in Benard convection*, published in Dynamical Chaos, Princeton University Press (1987), p. 71.

[6] MAJDA, A., Comm. Pure & Appl. Math. 39 (1986), pp. 187–220.

[7] ZAKHAROV, V.E., Sov. Phys. JETP, 35 (1972), 908.

[8] GOLDMAN, M., Rev. Mod. Phys. 56 (1984), 709–35.

[9] LANDMAN, M.J., PAPANICOLAOU, G.C., SULEM, C. & SULEM, P.L., Phys. Rev. A38, 3837–43 (1988). K. Rypdal, I. Rasmussen & K. Thomsen; Physica 16D, (1985), 339. I. Rasmussen & K. Rypdal; Phys. Scr. 33, 481 (1986).

[10] BARTUCCELLI, M., CONSTANTIN, P., DOERING, C., GIBBON, J.D. & GISSELFÄLT, M., Physica D, 44 (1990), 421–444.

[11] DOERING, C.D., GIBBON, J.D. & LEVERMORE, D., *Weak and strong solutions of the CGL equation*, preprint (1991).

[12] LEVERMORE, C.D., private communication.

[13] LANGE, C. & NEWELL, A.C., SIAM J. Appl. Math. 27 (1974), 441–56.

[14] KEEFE, L., Stud. Appl. Math., 73 (1985), 91.

[15] SIROVICH, L. & RODRIGUEZ, J., Phys. Lett. A120 (1988), 211.

[16] COULLET, P. GIL L. & LEGA, J., Phys. Rev Letts, 62 (1989), 1619–22.

[17] DOERING, C., GIBBON, J.D., HOLM, D.D. & NICOLAENKO, B., Nonlinearlity 1 (1988), 279–309 and Phys. Rev. Letts. 59 (1987), 2911–4.

[18] CONSTANTIN, P., *A construction of inertial manifolds*, Contemporary Mathematics 99, 27, (1989) (Proceedings AMS Summer School Boulder 1987).

[19] GHIDAGLIA, J–M. & HERON, B., 28D (1987), 282.

[20] HESLOT, F. CASTAING, B. & LIBCHABER, A., Phys. Rev. A36 (1987), 5870–3.

[21] DOERING, C.D., GIBBON, J.D., HOLM, D.D., HYMAN, J.M., LEVERMORE, D. & KOVACIC, G., in preparation (1990).

[22] BARTUCCELLI, M., DOERING, C.D., & GIBBON, J.D., *Ladder theorems for the 2D & 3D Navier Stokes equations on a finite periodic domain*, to appear in Nonlinearity 1991.

[23] TEMAM, R., *The Navier Stokes equations & nonlinear functional analysis*, CBMS–NSF Regional Conference Series in Appl. Math., SIAM 1983.

[24] HENSHAW, W.D., KREISS, H.O. & REYNA, L.G., *On the smallest scale for the incompressible Navier Stokes equations*, J. Theor. Appl. Fluid Mech., 1, (1989), 1–32. Henshaw, W.D., Kreiss, H.O. & Reyna, L.G. *Smallest scale estimates for the incompressible Navier Stokes equations*; preprint 1990.

[25] C.FOIAS AND R. TEMAM, *Gevrey class regularity for the solutions of the Navier Stokes equations*, J. of Funct. Anal. 87 (1989), 359–69.

SYMMETRIES, HETEROCLINIC CYCLES AND INTERMITTENCY IN FLUID FLOW*

PHILIP HOLMES†

Abstract. I review recent work, in which it is shown that the physical symmetries of certain fluid flows lead to equivariant dynamical systems that possess interesting global dynamics, in particular modulated travelling waves and heteroclinic cycles. The latter correspond to flow regimes in which the fluid remains relatively quiescent for "long" periods, punctuated by short, violent events in which other (higher) wavenumbers become active. These share many features of the bursting phenomenon in fully developed turbulent boundary layers and of intermittent events observed in other flows.

1. Introduction and a little history. Many turbulent flows exhibit intermittency in space and time. This occurs both in transition and in fully developed turbulence; in the latter the overall picture is very complicated, involving many length and time scales, but some aspects of transition and of the low wave-number turbulence production process in flows driven by shear may be amenable to study by dynamical systems methods. In this brief note I review some recent evidence in support of this; in particular I describe the rôle of homoclinic orbits and heteroclinic cycles and their relation to instabilities and the turbulence production mechanism at relatively low wavenumbers. High wavenumber spatio-temporal intermittency in the dissipation range is beyond the scope of this note and probably beyond the scope of "simple" ideas from dynamical system theory such as those put forward here.

I should make it clear that, while the heteroclinic cycles described here have been proven to occur in certain ODE's and PDE's, rigorous connections with Navier-Stokes in particular and turbulence in general are lacking. What rigorous results there are rely on local (center-unstable) manifold theorems and unfolding of degenerate bifurcations with symmetry (eg. Armbruster et al. [1988, 1989]) and as such are applicable at best to hydrodynamic instability and the transition to turbulence. Inertial manifolds, discussed at length in other papers in this volume, may eventually permit tighter links to "fully developed" turbulence production, but there is still some way to go. In spite of this, I feel that there is sufficient formal analytical and numerical evidence to suggest a connection between heteroclinic cycles in phase space and some aspects of intermittent turbulence production.

Busse and Heikes [1980] (Busse [1981]) were apparently the first to point out a connection between cycles and intermittent pattern rearrangment in Bénard convection in a thin, three dimensional layer. A three-dimensional ODE, derived formally to describe the interaction among the amplitudes of roll patterns oriented 120° apart, was found to exhibit a cycle consisting of three "pure mode" saddle points

---

*Supported by AFOSR 89-0226A (Wall Layers). Parts of the work described here were variously funded by ARO, NSF and ONR.

†Departments of Theoretical and Applied Mechanics and Mathematics, Cornell University, Ithaca, New York 14853.

connected by heteroclinic orbits. Busse and Heikes pointed out the importance of small stochastic perturbations in producing a variable period "statistical limit cycle." Guckenheimer and Holmes [1988] subsequently showed that this structure, which is due to equivariance of the ODE under cyclic permutation of the the the dependent variables, persisted under *all* small perturbations respecting the same symmetry: it was perhaps the first explicit example of a *structurally stable heteroclinic cycle.*

When such a cycle is attractive, solutions in its neighborhood spend increasingly long quiescent periods near the equilibria, punctuated by rapid heteroclinic transits. Numerical solutions of the Kuramoto-Sivaskinsky equation with "free" periodic boundary conditions due to Hyman et al. [1985, 1986] (Nicolaenko et al. [1985, 1986]) had earlier been observed to exhibit such behavior, but at that time the rôle of symmetry groups was not entirely clear. In this and in the ODE's derived by Galerkin projection of Navier-Stokes on an empirical eigenspace due to Aubry et al. [1988, 1990], cf. Holmes [1990], the appropriate group is $O(2)$, and Armbruster et al. [1988] were able to prove that this continuous group also led to structurally stable cycles in certain two mode interactions. Subsequently they showed that such an interaction accounted for the Kuramoto-Sivashinsky observations at "low" length parameters: Armbruster et al. [1989] also see Kevrekidis et al. [1990].

There has been a great deal of work on homoclinic and heteroclinic bifurcations in ODE's, most of it concerned with unfolding of degenerate (codimension ≥ 2) points, such as the Takens-Bogdanov points (Guckenheimer-Holmes [1983]). Here we are more concerned with large amplitude, more truly global situations in which existence and stability proofs do not depend upon locality in phase space. This work is more in the spirit of Silnikov [1965, 1968], plus symmetry. Recent contributions to the study of global symmetric cycles in ODE's includes Proctor and Jones' [1988] asymptotic analysis of $O(2)$-equivariant cycles in a convection problem (cf. Jones and Proctor [1987]), Knobloch and Moore's [1990], similar study of a doubly diffusive problem, Armbruster and Chossat's [1990] work on $O(3)$-symmetric systems of spherical harmonics, analyzing Friedrich and Haken's [1986] earlier numerical studies, Armbruster's [1989] preliminary work on that problem and on $O(2)$-cycles in multimode interactions, and more recent work on the latter of Campbell and Holmes [1990]. Also see Field and Swift [1990] and Melbourne et al. [1990]. Stone and Holmes [1988, 1990, 1991] have analyzed the effects of random and deterministic perturbations on such cycles and further evidence for their occurrence (of a mainly numerical nature) in boundary layer models including up to 64 modes have been provided by Aubry and Sanghi [1989, 1990] and Aubry [1990]. These latter papers concern systems equivariant under the group $O(2) \times SO(2)$. Larger scale numerical simulations of PDE's include those of Hyman et al. on the Kuramoto-Sivashinsky equations mentioned above as well as recent work of Nicolaenko and She [1990 a,b] an two-dimensional Kolmogorov flows: these latter clearly suggest a bursting mechanism based on $SO(2) \times D_k$-equivariant cycles to nonstationary solutions. Aubry et al. [unpublished] also observed cycles to periodic or quasiperiodic limit sets and Aubry and Sanghi [1990] contains illustrations of heteroclinic cycles to apparently "strange" limit sets. Somewhat more metaphysical evidence is to be found in Alan

Newell's "homoclinic orbits to infinity" in Langmuir turbulence (Newell et al. [1988 a,b] and in M. Shelly's computations on the one-dimensional complex Ginzburg-Landau equation in a "hard" régime in the sense of Bartuccelli et al. [1990], (J.D. Gibbon [these proceedings]). Finally, Leibovich and Mahalov [1990] have formal and numerical evidence of heteroclinic connections in several $SO(2)$-equivariant systems modelling wave interactions among invariant subspaces of Navier-Stokes in circular and plane geometries, with implications for transition in pipe flows (rotating Hagen-Poiseuille flow) and boundary layers. Mullin and Darbyshire [1990] have striking phase space reconstructions which demonstrate heteroclinic and homoclinic cycles of Silnikov type in an $SO(2)$ equivariant, short Taylor-Coutte apparatus with rotating endplates. However, these cycles most likely derive from a degenerate codimension two difurcation and do not appear to be structurally stable and it is not clear how the continuous and discrete symmetries of their problem are involved. This list is certainly incomplete, even at the time of writing.

In the remainder of this note I describe a specific example in complex two-space (that of Armbruster et al. [1988]) which motivates the more general remarks that follow.

2. Symmetries in physical and phase space: An example. Suppose that

$$(2.1) \qquad \frac{du}{dt} = Au + F(u)$$

denotes a partial (integro-) differential equation in abstract form. Let $u = u(x, t)$ depend on a single space dimension, for simplicity, and suppose that (2.1) is equivariant with respect to translation $x \to x + \alpha$ and reflection $x \to -x$. This occurs, for example, if the differential operators in A and F are of even order or occur in even functions of odd order operators, as in the Kuramoto-Sivashinsky equation in the form

$$(2.2) \qquad u_t + u_{xxxx} + u_{xx} + \frac{1}{2}(u_x)^2 = 0.$$

"Free" periodic boundary conditions are also required, for infinitesimal translation invariance. If the solutions of (2.1) are represented by Fourier modes

$$(2.3) \qquad u(x, t) = \sum_{-\infty}^{\infty} a_k(t) e^{ikx}$$

(with a suitable reality condition, if necessary) then the resulting ODE's

$$(2.4) \qquad \dot{a}_k = f(a_1, a_2, \cdots, a_k, \cdots) \quad k = 1, 2, \cdots$$

will be equivariant under the actions

$$(2.5a,b) \qquad a_k \longmapsto a_k e^{ik\alpha} \text{ and } a_k \longmapsto a_k^*.$$

This is representation of $O(2)$: the group of planar reflections and rotations.

The interaction between two modes with spatial wavenumbers in the ratio 1:2 can be written in $0(2)$ equivariant normal form as

$$(2.6) \qquad \left. \begin{aligned} \dot{a}_1 &= a_1^* a_2 + \left(\mu_1 + e_{11} |a_1|^2 + e_{12} |a_2|^2 \right) a_1 \\ \dot{a}_2 &= \pm a_1^2 + \left(\mu_2 + e_{21} |a_1|^2 + e_{21} |a_2|^2 \right) a_2 \end{aligned} \right\} + \mathcal{O}\left(|a|^4 \right)$$

(Dangelmayr and Armbruster [1986]). Note that invariance under complex conjugation (2.5b) (reflection) implies that the real subspace is invariant for (2.6). For certain values of the real parameters μ_j, e_{jk}, the phase portrait of this reduced system reveals a pair of orbits connecting a saddle point $A : Re(a_2) = +\sqrt{\frac{-\mu_2}{e_{22}}}$ to a sink $B : Re(a_2) = -\sqrt{\frac{-\mu_2}{e_{22}}}$. However, in the full phase space $\mathbb{C}^2$ of (2.6) these equilibria belong to a circle $S : |a_2| = \sqrt{\frac{-\mu_2}{e_{22}}}$ of equilibria (an orbit of the rotation group (2.5a)) and each one has a one dimensional unstable manifold and a two dimensional stable manifold in addition to the neutral (center) manifold along the circle itself. Application of the group element $a_k \longmapsto a_k e^{\frac{ik\pi}{2}}$ takes the connection $A \longmapsto B$ in the $(Re(a_1), Re(a_2))$ plane into a connection $B \longmapsto A$ in the $(Im(a_1), Re(a_2))$-plane, completing the cycle. See Figure 1. Of course, every diametrically opposite pair of equilibria on S is thus heteroclinically connected, since Figure 1 may be infinitesimally rotated by (2.5a).

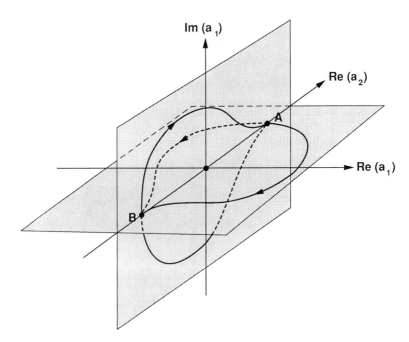

Figure 1. An $O(2)$ equivariant heteroclinic cycle in a four dimensional system ($Im(a_2)$ suppressed). See Armbruster et al. [1988].

Equivariant structural stability of the cycles follows from the fact that the real subspace and rotations of it under (2.5a) remain invariant under all $O(2)$-equivariant perturbations, and the saddle-sink connections in these subspaces are themselves structurally stable. Attractivity of the cycles is determined by the ratios of (the real parts of) certain "relevant" eigenvalues of the saddle points, much in the manner of Silnikov's [1965] original analysis of homoclinic orbits.

Reconstructions in physical space by (2.3) reveal that solutions attracted to such a cycle spend increasing periods near "frozen" spatial states (here the pure second Fourier mode) interspersed with transitions in which other modes (here only the first) grow rapidly and then decay. The signature of such a cyclic event involves a characteristic phase shift (here by $\alpha = \frac{\pi}{2}$ in (2.5a): π in mode 2, $\frac{\pi}{2}$ in mode 1). When more modes are active, as in Aubry et al. [1988, 1990] and Aubry and Sanghi [1989, 1990] the dynamics is correspondingly richer; in particular especially interesting régimes are observed in models of fluid flows in which the modes belonging to an invariant subspace having no streamwise (axial) variations oscillate in an apparently chaotic fashion while those having streamwise variations burst intermittently. (We do not yet understand this fully).

In the boundary layer application these heteroclinic events seem to correspond rather well to "bursts" (Kline et al. [1967], Kline [1978], cf. Blackwelder [1989]) in which the *relatively* steady (although far from laminar) wall layer flow an event suddenly erupts, ejecting fluid with low axial velocity into the upper layer, with a concomitant burst of Reynolds stress, followed by a "sweep" of high speed fluid into the wall region and subsequent reformation of low speed steaks, often shifted laterally with respect to those that preceded the burst. Figure 2 shows how the cross-stream components of the velocity field evolve during such a burst.

The other "heteroclinic scenarios" referred to in section one are broadly similar to that outlined above, although the correspondences between physical and phase space and the details vary widely. Of particular interest with respect to (two dimensional) turbulence are the computations of Nicolaenko and She [1990 a,b] on the Kolmogorov flow, who find that the heteroclinic cycles correspond to (generate?) a burst of enstrophy on the way out followed by an inverse cascade of energy back to low wavenumbers as the flow "relaminarizes" near a (probably) hyperbolic set corresponding to two large scale counter-rotating vortices.

3. Asymmetric, time dependent and noisy perturbations of heteroclinic cycles. In the ODE's modelling the near wall region of a boundary layer derived by Aubry et al. [1988], a small additive term deriving from pressure fluctuations at the upper edge of the domain (essentially a free surface) was initially neglected. This term represents communication–the weak forcing–from the outer region. After finding that attracting heteroclinic cycles occurred in the "unforced" model, the importance of this term became clear: it would significantly affect solutions passing near the unperturbed saddle points, and so would "randomize" the recurrence or passage times, without much affecting the structure of the bursts themselves in phase or physical space, for on the heteroclinic cycles far from equilibria the forcing term is relatively small. Analyses of Stone and Holmes [1989, 1990]

Time increasing

↓

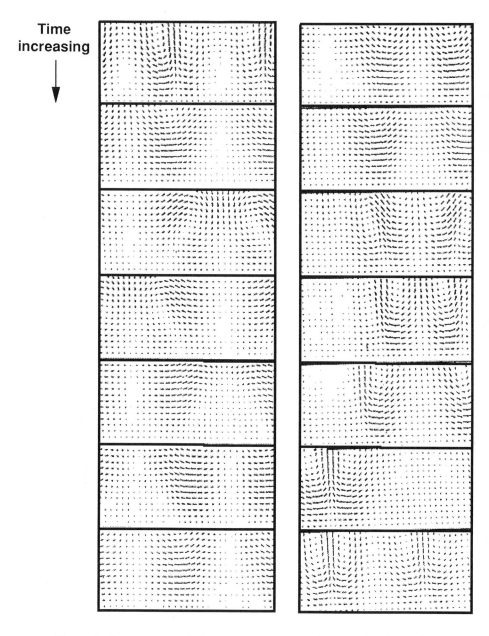

Figure 2. Cross-stream velocity components at equi-spaced time intervals and a fixed streamwise location in a reconstruction of the near wall boundary layer during a heteroclinic excursion in the associated dynamical system, from Aubry and Sanghi [1990].

using additive Wiener process forcing revealed a distribution of inter-event durations with a strong exponential tail, the exponent being simply the leading unstable eigenvalue at the saddle. Significantly, such distributions of inter-burst durations

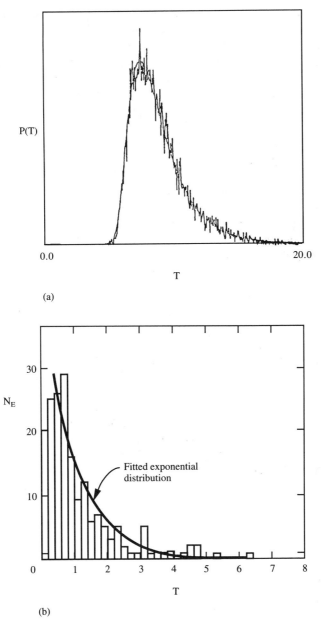

Figure 3. Distributions of inter-event durations. (a) From the theory of Stone and Holmes [1990] for randomly perturbed heteroclinic cycles (compared with numerical simulations). (b) Experimental determination of interburst durations in a turbulent boundary layer, from Bogard and Tiederman [1985]

have been measured in boundary layer experiments (eg. Kim et al. [1971], Bogard and Tiederman [1986]) and in other situations including spatially simpler instabilities in closed systems (Mullin and Darbyshire [1990]. See Figure 3. More recently, Stone and Holmes [1991] realized that stochastic forcing is not necessary: deterministic perturbations that yield an attractor with a reasonable invariant measure of virtually *any* shape will also give a strongly skewed distribution of return times having such an exponential tail (cf. Brunsden [1987, 1989], in which power spectra of such motions are also calculated).

This suggest that the precise form of the (weak) perturbation or forcing (or coupling to other modes) is unimportant: it merely serves to catalyze or release the dynamical behavior inherent to the symmetric system. In the boundary layer context this appears to have significant implications for the way in which bursting statistics scale in terms of "inner" or "outer" variables or a mixtures of both, (Aubry et al. [1988], Holmes [1990]). In a more general context one might imagine that such heteroclinic bursting behavior, induced by (approximate) symmetries of the physical problem, would be relatively insensitive to the details of eddy viscosity models or averaging and homogenization principles which one might apply to remove or replace shorter wavelength modes in the inertial and dissipative ranges.

A general principle. Notwithstanding all their omissions, the works reviewed above suggest a general principle: that turbulence production and instability mechanisms in which periods of relative calm with few excited modes are interrupted, irregularly, by storms of activity involving more spatial modes and higher wavenumber motions, correspond to orbits circulating in the neighborhood of heteroclinic cycles induced by the natural symmetries of the physical problem. If this picture is correct, it has important implications for control of such intermittent events (cf. Bloch and Marsden [1990], Berkooz et al. [1990]). Perhaps more significantly, it provides a unified global description, in phase space, of a local (linear) instability mechanism, its strongly nonlinear growth phase and the subsequent collapse back to the neighborhood of another unstable steady or quasi-steady solution. Perhaps it may help us understand the characteristic life cycle of coherent structures in flows such as the boundary layer (Kline [1978]).

REFERENCES

[1] D. ARMBRUSTER, *More on structurally stable H-orbits*, Proc. Int. Conf. on Bifurcation Theory and Its Numerical Analysis (ed. Li Kaitai, J. Marsden, M. Golubitsky and G. Iooss) (1989), pp. 1–13., Xian Jiaotong University Press, P.R.C.

[2] D. ARMBRUSTER AND P. CHOSSAT, *Heteroclinic orbits in a spherically invariant system.*, Univ. de Nice Prépublication, 259 (1990).

[3] D. ARMBRUSTER, J. GUCKENHEIMER AND P. HOLMES, *Heteroclinic cycles and modulated travelling waves in systems with O(2)-symmetry*, Physica, 29D (1988), pp. 257–282.

[4] D. ARMBRUSTER, J. GUCKENHEIMER AND P. HOLMES, *Kuramoto-Sivashinsky dynamics on the center-unstable manifold*, S.I.A.M. J. on Applied Math, 49 (1989), pp. 676–691.

[5] N. AUBRY, *Use of experimental data for an efficient description of turbulent flows*, Applied Mechanics Reviews, 43 5(2) (1990), pp. S240–S245.

[6] N. AUBRY, P. HOLMES, J. LUMLEY AND E. STONE, *The dynamics of coherent structures in the wall region of a turbulent boundary layer*, J. Fluid Mech, 192 (1988), pp. 115–173.

[7] N. AUBRY, J.L. LUMLEY AND P.J. HOLMES, *The effect of modeled drag reduction on the wall region*, Theoret. Comput Fluid Dynamics 1 (1990), pp. 229–248.

[8] N. AUBRY AND S. SANGHI, *Streamwise and spanwise dynamics of the turbulent wall layer*, In Forum on Chaotic Flow (ed. Ghia) (1989), ASME, New York.

[9] N. AUBRY AND S. SANGHI, *Bifurcation and bursting of streaks in the turbulent wall layer*, ed. M. Lesieur and O. Métais, In Turbulence 89 Organized Structure and Turbulence in Fluid Mechanics, Kluwer Academic Publishers, 1990.

[10] M. BARTUCCELLI, P. CONSTANTIN, C.R. DOERING, J.D. GIBBON AND M. GISSELFÄLT, *On the possibility of soft and hard turbulence in the complex Ginzburg Landau equation, (preprint)*.

[11] G. BERKOOZ, J. GUCKENHEIMER, P. HOLMES, J.L. LUMLEY, J.E. MARSDEN, N. AUBRY AND E. STONE, *Dynamical systems theory approach to the wall region, June 18-20, 1990*, Proc AIAA Fluid Dynamics, Plasma Dynamics and Lasers Conference, Seattle, Washington.

[12] R.F. BLACKWELDER, *Some ideas on the control of near wall eddies*, AIAA paper 89-1009, AIAA 2nd Shear Flow Conference (1989).

[13] A.M. BLOCH AND J.E. MARSDEN, *Controlling homoclinic orbits*, Theor. Comp. Fluid Dyn., 1 (1990), pp. 179–190.

[14] V. BRUNSDEN, J. CORTELL AND P. HOLMES, *Power spectra of chaotic vibrations of a buckled beam*, J. Sound Vib., 130 (1989), pp. 1–25.

[15] V. BRUNSDEN AND P. HOLMES, *Power spectra of strange attractors near homoclinic orbits*, Phys. Rev. Lett, 58 (1987), pp. 1699–1702.

[16] F.M. BUSSE AND K.E. HEIKES, *Convection in a rotating layer: A simple case of turbulence*, Science, 208 (1980), pp. 173–175.

[17] F.M. BUSSE, *Transition to turbulence in Rayleigh-Bénard convection*, Hydrodynamic Instabilities and the transition to turbulence (ed. H.L Swinney and J. P. Gollub) (1981), pp. 97–137, Springer-Verlag.

[18] S. CAMPBELL AND P. HOLMES, *(submitted) Bifurcation from O(2)-Symmetric heteroclinic cycles with three interacting modes*.

[19] G. DANGELMAYR AND D. ARMBRUSTER, *Steady state mode interactions in the presence of O(2)-symmetry and in non-flux boundary value problems.*, Multiparameter Bifuraction Theory (ed. M. Golubitsky and J. Guckenheimer), Cont. Math., 56 (1986), pp. 53–68.

[20] M FIELD AND J.W. SWIFT, *Stationary bifurcation to limit cycles and heteroclinic cycles*, MSI Technical Report, 90-29, Cornell University.

[21] R. FRIEDRICH AND H. HAKEN, *Static, wavelike and chaotic thermal convection in spherical geometries*, Phys. Rev. A., 34 (1986), pp. 2100–2120.

[22] M. GOLUBITSKY AND J. GUCKENHEIMER (EDS.), *Multiparameter Bifurcation Theory*, AMS Contemporary Mathematics, 56, Providence, Rhode Island.

[23] J. GUCKENHEIMER AND P. HOLMES, *Nonlinear Oscillations, Dynamical Systems and Bifurcations of Vector Fields*, (Corrected third printing, 1990), Springer Verlag, New York.

[24] J. GUCKENHEIMER AND P. HOLMES, *Structurally stable heteroclinic cycles*, Math. Proc. Camb. Phil. Soc., 103 (1988), pp. 189–192.

[25] P.J. HOLMES, *Can dynamical systems approach turbulence?*, In Whither Turbulence? Turbulence at the Crossroads (ed. J.L. Lumley), Springer Lecture Notes in Physics, 357 (1990), pp. 197–249 and 306–309, Springer Verlag, New York.

[26] J.M. HYMAN AND B. NICOLAENKO, *The Kuramoto-Sivashinsky equation: A bridge between PDE's and dynamical systems*, Los Alamos National, Lab Report LA-UR-85-1556 (1985).

[27] J.M. HYMAN, B. NICOLAENKO AND S. ZALESKI, *Order and complexity in the Ku-ramoto-Sivashinsky model of weakly turbulent interfaces.*, Los Alamos National, Lab Report LA-UR-86-1947 (1986).

[28] C. JONES AND M.R. PROCTOR, *Strong spatial resonances and travelling waves in Bénard convection*, Phys. Lett, A 121 (1987), pp. 224-227.

[29] I.G. KEVREKIDIS, B. NICOLAENKO AND J.C. SCOVEL, *Back in the saddle again: A computer assisted study of the Kuramoto-Sivashinky equation*, S.I.A.M. J. on Applied Math, 50 (1990), pp. 760-790.

[30] H.T. KIM, S.J. KLINE AND W.C. REYNOLDS, *The production of turbulence near a smooth wall in a turbulent boundary layer*, J. Fluid Mech., 50 (1971), pp. 133-160.

[31] S.J. KLINE, *The role of visualization in the study of the turbulent boundary layer*, Proc AFOSR/Lehigh Workshop, ed. C.R. Smith and D.E. Abbott, Coherent Structure of Turbulent Boundary Layers (1978).

[32] S.J. KLINE, W.C. REYNOLDS, F.A. SCHRAUB AND P.W. RUNDSTATLER, *The structure of turbulent boundary layers*, J. Fluid Mech., 30 (1967), pp. 741-773.

[33] S. LEIBOVICH AND A. MAHALOV, *Resonant interactions in rotating pipe flow*, (submitted to J. Fluid Mech.) (1990).

[34] I. MELBOURNE, *Intermittency as a codimension three phenomenon*, J. Dyn. and Diff. Eqns., 1 (1989), pp. 347-367.

[35] I. MELBORNE P. CHOSSAT AND M. GOLUBITSKY, *Heteroclinic cycles involving periodic solutions in mode interactions with* $O(2)$ *symmetry*, Proc. R. Soc. Edinburgh (in press) (1990).

[36] H.K. MOFFATT, *Fixed points of turbulent dynamical systems and suppression of nonlinearity*, Whither Turbulence: Turbulence at the Crossroads, (ed. J.L Lumley), Springer Lecture Notes in Physics, 357 (1990), pp. 250-257, Springer Verlag, New York.

[37] T. MULLIN AND A.G. DARBYSHIRE, *Intermittency in a rotating annular flow*, (submitted to Europhys Lett A) (1990).

[38] A.C. NEWELL, D.A.RAND AND D. RUSSELL, *Turbulent dissipation rates and the random occurrence of coherent events*, Phys. Lett A, 132 (1988a), pp. 112-123.

[39] A.C. NEWELL, D.A. RAND AND D. RUSSELL, *Turbulent transport and the random occurrence of coherent events*, Physica, 33 D (1988b), pp. 281-303.

[40] B. NICOLAENKO, B. SCHEURER AND R. TEMAM, *Some global dynamical properties of the Kuramoto-Sivashinsky equations: nonlinear stability and attractors*, Physica, 16 D (1985), pp. 155-183.

[41] B. NICOLAENKO, B. SCHEURER AND R. TEMAM, *Attractors for the Kuramoto-Sivashinsky equations*, Los Alamos National Lab, Report LA-UR-85-1630 (1986).

[42] B. NICOLAENKO AND A.S. SHE, *Temporal intermittency and turbulence production in the Kolmogorov flow*, Topological Dynamics of Turbulence (1990a) (to appear), Cambridge University Press.

[43] B. NICOLAENKO AND A.S. SHE, *Symmetry-breaking homoclinic chaos in Komogorov flows*, preprint (1990b), Arizona State University).

[44] M.K. PROCTOR AND C. JONES, *The interaction of two spatially resonant patterns in thermal convection I: Exact 1:2 resonance*, J. Fluid Mech., 188 (1988), pp. 301-335.

[45] L.P. SILNIKOV, *A case of the existence of a denumerable set of periodic motions*, Sov. Math. Dokl., 6 (1965), pp. 163-166.

[46] L.P. ŠILNIKOV, *On the generation of a periodic motion from trajectories doubly asymptotic to an equilibrium state of saddle type*, Math. U.S.S.R. Sbornik, 6 (1968), pp. 427-438.

[47] E. STONE AND P. HOLMES, *Noise induced intermittency in a model of a turbulent boundary layer*, Physica, D37 (1989), pp. 20-32.

[48] E.STONE AND P. HOLMES, *Random perturbations of heteroclinic attractors*, SIAM J. on Applied Math, 50 (1990), pp. 726-743.

[49] E.STONE AND P. HOLMES, *Unstable fixed points, homoclinic cycles and exponential tails in turbulence*, Proc. Symposium in Honor of J. L. Lumley (1991) (to appear).

FINITE-DIMENSIONAL DESCRIPTION OF
DOUBLY DIFFUSIVE CONVECTION

E. KNOBLOCH*, M.R.E. PROCTOR** AND N.O. WEISS**

Abstract. Doubly diffusive convection in small aspect ratio systems exhibits complex temporal dynamics that has been attributed to the Shil'nikov mechanism. These results are reviewed and an asymptotic expansion suggested that leads in a systematic manner from the partial differential equations to a third order system of ordinary differential equations with Shil'nikov dynamics.

0. Introduction. The purpose of this article is to describe some attempts at providing a finite-dimensional description of the complex behavior that is known to occur in doubly diffusive systems. In section I we summarize the results of numerical experiments performed on the partial differential equations describing thermosolutal convection in small aspect ratio systems. In section II we describe an asymptotic régime in which these equations can be reduced systematically to a third order system of ordinary differential equations. This system contains the dynamics first described by Shil'nikov [22] and elaborated by Glendinning and Sparrow [23]. This dynamics bears a strong qualitative resemblance to the numerical results, suggesting that the Shil'nikov mechanism is responsible for the observed dynamics even for parameter values outside the asymptotic régime. Support for this conjecture is provided by truncations of (flat) Galerkin expansions which also exhibit the Shil'nikov mechanism. In section III we describe how our results are modified in large aspect ratio systems (modelled here by imposing periodic boundary conditions in the horizontal) and discuss prospects for improved finite-dimensional models of doubly diffusive systems.

I. Doubly diffusive convection. Doubly diffusive systems are characterized by a competition between a destabilizing force (typically thermal buoyancy) and a stabilizing force. The stabilization may be provided by an ambient concentration gradient as in thermosolutal convection; in binary mixtures the concentration gradient is set up in response to the thermal forcing by means of the Soret effect. In magnetoconvection or in rotating systems the restoring force is provided by the Lorentz and Coriolis forces, respectively. These systems differ from Rayleigh-Bénard convection in that the primary instability may be oscillatory. In thermosolutal convection such an instability occurs when during an oscillation the phase lag between the thermal and concentration fields extracts sufficient energy from the thermal stratification to overcome viscous dissipation. The resulting instability is diffusive and does not require the system to be dynamically unstable.

---

*Department of Physics, University of California, Berkeley CA 94720, USA.

**Department of Applied Mathematics and Theoretical Physics, University of Cambridge, Cambridge CB3 9EW, UK.

(a) The equations.

The basic equations describing thermosolutal convection are the Boussinesq equations

(1a)
$$\rho_0 \left(\frac{\partial \mathbf{u}}{\partial t} + \mathbf{u} \cdot \nabla \mathbf{u} \right) = -\nabla p + \rho \mathbf{g} + \rho_0 \nu \nabla^2 \mathbf{u}$$

(1b)
$$\frac{\partial T}{\partial t} + \mathbf{u} \cdot \nabla T = \kappa_T \nabla^2 T$$

(1c)
$$\frac{\partial S}{\partial t} + \mathbf{u} \cdot \nabla S = \kappa_S \nabla^2 S$$

with

(1d)
$$\nabla \cdot \mathbf{u} = 0, \quad \rho = \rho_0 (1 - \alpha T + \beta S).$$

Thus the Navier-Stokes equation is coupled to two advection-diffusion equations through the buoyancy term $\rho \mathbf{g}$. Here $\mathbf{u}, p, T$ and S are the velocity, pressure, temperature and concentration fields, ρ is the density, and ν, κ_T and κ_S are the kinematic viscosity, thermal diffusivity and solutal diffusivity, respectively, assumed to be constant. The coefficients of expansion α, β are positive constants.

In the simplest situation we assume that the boundary conditions at the top and bottom of a plane layer are stress-free and that the temperature and concentration are fixed there:

(2a)
$$\frac{\partial u}{\partial z} = \frac{\partial v}{\partial z} = w = 0, \; T = T_0 + \Delta T, \; S = S_0 + \Delta S \quad \text{on } z = 0$$

(2b)
$$\frac{\partial u}{\partial z} = \frac{\partial v}{\partial z} = w = 0, \; T = T_0, \qquad S = S_0 \qquad \text{on } z = Hh,$$

where $\mathbf{u} = (u, v, w)$, $\Delta T > 0$, $\Delta S > 0$, and H is dimensionless. It is convenient to define quantities θ, ϕ by

(3)
$$T = T_0 + \Delta T \left(1 - \frac{z}{Hh} + \theta \right), \quad S = S_0 + \Delta S \left(1 - \frac{z}{Hh} + \phi \right),$$

so that θ, ϕ denote departures of the temperature and concentration from the no-motion-state

(4)
$$\mathbf{u} = 0, \quad T = T_0 + \Delta T \left(1 - \frac{z}{Hh} \right), \quad S = S_0 + \Delta S \left(1 - \frac{z}{Hh} \right)$$

(hereafter the trivial solution). In two dimensions one may introduce the dimensionless streamfunction $\psi(x, z, t)$ such that $\mathbf{u} \equiv \kappa_T(-\partial_z \psi, 0, \partial_x \psi)$. In terms of the dimensionless variables $x/h, z/h, t/(h^2/\kappa_T)$ equations (1) become

(5a)
$$\sigma^{-1} \left[\partial_t \nabla^2 \psi + J(\psi, \nabla^2 \psi) \right] = R_T \partial_x \theta - R_S \partial_x \phi + \nabla^4 \psi$$

(5b)
$$\partial_t \theta + J(\psi, \theta) = \partial_x \psi + \nabla^2 \theta$$

(5c)
$$\partial_t \phi + J(\psi, \phi) = \partial_x \psi + \tau \nabla^2 \phi,$$

where $J(f,g) \equiv (\partial_x f)(\partial_z g) - (\partial_z f)(\partial_x g)$, and the dimensionless parameters are given by

$$(5d) \qquad \sigma = \frac{\nu}{\kappa_T}, \quad \tau = \frac{\kappa_S}{\kappa_T}, \quad R_T = \frac{g\alpha\Delta T h^3}{\kappa_T \nu}, \quad R_S = \frac{g\beta\Delta S h^3}{\kappa_T \nu}.$$

The corresponding boundary conditions are

$$(5e) \qquad \psi = \psi_{zz} = \theta = \phi = 0 \quad \text{on} \quad z = 0, H.$$

A Hopf bifurcation from the trivial solution $\psi = \theta = \phi = 0$ occurs if $\tau < 1$ and $R_S > R_{S,CT}$ (see below).

(b) Numerical results.

Detailed numerical studies of equations (5) are available [7,13,17,18]. These calculations are carried out with the lateral boundary conditions

$$(5f) \qquad \psi = \psi_{xx} = \theta_x = \phi_x = 0 \quad \text{on} \quad x = 0, \Lambda.$$

These boundary conditions force the planes $x = 0$, $x = \Lambda$ to be stress-free but impenetrable walls. The computations use $\Lambda = 2^{1/2}$ or 1.5 corresponding (approximately) to half the wavelength of the mode that first loses stability with increasing Rayleigh number R_T. The computations impose the symmetry

$$(6) \quad \psi(x,z) = \psi(\Lambda - x, H - z), \theta(x,z) = -\theta(\Lambda - x, H - z), \phi(x,z) = -\phi(\Lambda - x, H - z)$$

and employ a second order accurate finite difference scheme. The mesh typically used is equivalent to 594 independent variables. The dependence of the results on mesh size and time step has been extensively investigated [18]. In Figure 1 we summarize the results of these computations. The figure shows the Nusselt number, related to the square of the amplitude of motion, as a function of R_T. The oscillations set in at $R_T^{(o)}$, increase in amplitude and undergo a sequence of bifurcations, the first of which breaks the *temporal* symmetry

$$(7) \quad \begin{aligned} &\psi(x,z,t) = -\psi\left(x, H - z, t + \frac{1}{2}P\right), \quad \theta(x,z,t) = -\theta\left(x, H - z, t + \frac{1}{2}P\right), \\ &\phi(x,z,t) = -\phi\left(x, H - z, t + \frac{1}{2}P\right), \end{aligned}$$

where P is the oscillation period. We refer to the oscillations with the symmetry (7) as symmetric and those that break the symmetry (7) as asymmetric. Both types of solutions retain the imposed *spatial* symmetry (6). The asymmetric oscillations undergo a cascade of period-doubling bifurcations into chaos and back out of it. The chaotic region is interspersed with windows containing more complicated periodic oscillations, both symmetric and asymmetric, which themselves undergo period doubling bifurcations. For the symmetric oscillations these are again preceded by a bifurcation to asymmetry. The whole structure is repeated hysteretically in a

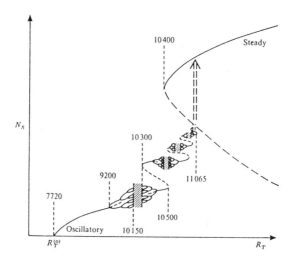

Figure 1. Schematic bifurcation diagram for numerical experiments on two-dimensional thermosolutal convection, showing both oscillatory and steady branches. Conjectured unstable solutions are denoted by broken lines. (From Knobloch et al [13].)

sequence of bifurcation "bubbles", four of which are shown in Figure 1. The results provide the most extensive study of chaotic behavior in a hydrodynamic system described by partial differential equations.

(c) The Shil'nikov mechanism.

Moore et al [17] suggested that the above behavior is related to the presence of a heteroclinic orbit connecting two saddle-foci to one another. These saddle-foci correspond to unstable steady (overturning) convection (hereafter SS) and are characterised by their eigenvalues. Suppose that the three dominant (i.e., least stable) eigenvalues are

(8a) $$\gamma, -\alpha \pm i\beta, \quad \gamma > 0, \ \alpha > 0.$$

Since the saddle-foci are related by reflection in $x = \Lambda/2$ their eigenvalues are identical. We define

(8b) $$\delta \equiv \alpha/\gamma$$

and let δ_h denote the value of δ at the parameter value for which a global connection is present. Shil'nikov [22] showed for a homoclinic connection that the condition $\delta_h < 1$ implies the presence of a countable number of horseshoes in the dynamics of the system. Motivated by the results of Figure 1 Glendinning and Sparrow [5] investigated in detail the sequence of bifurcations that gives rise to the complex orbits guaranteed by Shil'nikov's theorem as a bifurcation parameter μ passes through

μ_h. Their results are summarized in Figure 2 showing the period $P(\mu)$ for the two cases $\delta > 1$ and $\frac{1}{2} < \delta < 1$. In the former the period increases monotonically as $\mu \to \mu_h$. In the latter the infinite period is approached by a series of wiggles containing period-doubling bubbles with chaotic intervals and hysteresis between successive bubbles. In addition their study revealed the existence of subsidiary homoclinic orbits with their own associated bubble structure. Since the period and amplitude are (typically) in 1-1 correspondence the analysis leads to results much like those in Figure 1. The above structure is present also for $\delta < \frac{1}{2}$ but is unstable.

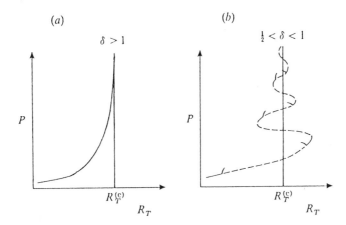

Figure 2. Period P as a function of the bifurcation parameter μ in the neighborhood of the homoclinic bifurcation when (a) $\delta > 1$, (b) $1/2 < \delta < 1$. Full and broken lines represent stable and non-stable solutions, respectively.

The presence of a heteroclinic orbit in the partial differential equations is difficult to establish. Only in the limit where the two primary bifurcations coalesce ($R_T^{(o)} = R_T^{(e)}$), i.e.,

(9a) $$R_T = R_{T,CT} \equiv \frac{\sigma + \tau}{\sigma(1 - \tau)} R_0$$

(9b) $$R_S = R_{S,CT} \equiv \frac{\tau^2}{\sigma} \left(\frac{1 + \sigma}{1 - \tau} \right) R_0$$

(9c) $$R_0 \equiv \pi^4 (H^2 + \Lambda^2)^3 / \Lambda^4 H^6,$$

and the dynamics of the p.d.e.'s reduce to the Takens-Bogdanov normal form

(10a) $$\dot{v} = w$$

(10b) $$\dot{w} = \mu v + \nu w + A v^3 + B v^2 w$$

with $A > 0, B < 0$, and $|\mu|, |\nu| \ll 1$ can one show the existence of a heteroclinic connection between the two saddle-points [6,14]. In this limit the saddle points are characterized by two dominant eigenvalues, one small and positive and one small and negative. All the other eigenvalues are $O(1)$ and negative. Consequently the vector field (10) is planar and no interesting dynamics occur near such a connection. Moore et al [17] and Knobloch et al [13] pointed out, however, that with increasing $R_T - R_{T,CT}$ and $R_S - R_{S,CT}$ the two least negative eigenvalues coalesce on the negative real axis, and thereafter move into the complex plane giving rise to the situation (8a). This scenario is supported by studying truncations of flat Galerkin expansions of which the simplest nontrivial one is

(11a) $$\dot{a} = \sigma[-a + rb - sd]$$

(11b) $$\dot{b} = -b + a(1 - c)$$

(11c) $$\dot{c} = \varpi[-c + ab]$$

(11d) $$\dot{d} = -\tau d + a(1 - c)$$

(11e) $$\dot{e} = \varpi[-\tau e + ad],$$

where $r \equiv R_T/R_0, S \equiv R_S/R_0$ and $\varpi \equiv 4\Lambda^2/(H^2 + \Lambda^2)$. Numerical study of (11) indicates the presence of the Shil'nikov mechanism but in a régime where higher order modes can no longer be neglected [1,16]. In the following section we describe a systematic asymptotic procedure that enables us to derive a set of ordinary differential equations possessing the Shil'nikov mechanism. The basic idea is to seek a codimension-three singularity that would add another equation to the Takens-Bogdanov normal form (10) and so describe the deformation of the planar heteroclinic orbit into one joining the saddle-foci. Since in the present problem there are no other primary bifurcations we look at the limit $\Lambda \to 0$, i.e., study convection in tall thin cells.

II. The asymptotic régime.

Before proceeding it is convenient to define horizontally averaged quantities denoted by an overbar, and fluctuating quantities denoted by a tilde. Thus, for example, $\bar{\theta} \equiv \frac{1}{\Lambda} \int_0^\Lambda \theta dx$ while $\tilde{\theta} \equiv \theta - \bar{\theta}$. To look at tall thin cells we set $\Lambda = \pi$ and let H be large. Following Proctor and Weiss [21] we use $\epsilon \equiv \pi H^{-1}$ as an expansion parameter and let

(12a) $$\zeta = \epsilon z, \quad T = \epsilon^2 t,$$

$$\psi = \epsilon^2 \tilde{\psi}(x, \zeta, T; \epsilon), \quad \theta = \epsilon^2 \tilde{\theta}(x, \zeta, T; \epsilon) + \epsilon^3 \bar{\theta}(\zeta, T; \epsilon),$$

(12b) $$\phi = \epsilon^2 \tilde{\phi}(x, \zeta, T; \epsilon) + \epsilon^3 \bar{\phi}(\zeta, T; \epsilon),$$

where

(12c) $$\tilde{\psi} = \tilde{\psi}_0 + \epsilon^2 \tilde{\psi}_2 + \cdots, \tilde{\theta} = \tilde{\theta}_0 + \epsilon^2 \tilde{\theta}_2 + \cdots, \tilde{\phi} = \tilde{\phi}_0 + \epsilon^2 \tilde{\phi}_2 + \cdots$$

(12d) $$\bar{\theta} = \bar{\theta}_0 + \epsilon^2 \bar{\theta}_2 + \cdots, \quad \bar{\phi} = \bar{\phi}_0 + \epsilon^2 \bar{\phi}_2 + \cdots.$$

In addition we expand R_T and R_S around the codimension-two point:

(12e) $$R_T = R_{T,CT} + \epsilon^2 R_{T,2} + \epsilon^4 R_{T,4} + \cdots$$

(12f) $$R_S = R_{S,CT} + \epsilon^2 R_{S,2} + \epsilon^4 R_{S,4} + \cdots.$$

Here, as in (7), the overbar indicates a horizontal average and the tilde the fluctuating part. At $O(\epsilon^0)$ we obtain from (5) the solution

$$(13) \qquad \tilde{\psi}_0 = a \sin x, \quad \tilde{\theta}_0 = \tau \tilde{S}_0 = a \cos x.$$

The ζ-dependence of $a(\zeta, T)$ follows from a solvability condition at $O(\epsilon^2)$:

$$(14) \qquad a = \hat{a}(T) \sin \zeta, \quad R_{T,2} - R_{S,2}/\tau = 3.$$

The solution of the $O(\epsilon^2)$ problem is then given by

$$(15a) \qquad \tilde{\psi}_2 = b_\psi \sin x, \quad \tilde{\theta}_2 = b_\theta \cos x, \quad \tilde{\phi}_2 = \frac{b_\phi}{\tau} \cos x,$$

where

$$(15b) \quad b_\psi \equiv b(\zeta, T), \quad b_\theta \equiv -\left(\frac{\partial \hat{a}}{\partial T} + \hat{a}\right) + b(\zeta, T), \quad b_\phi \equiv -\left(\frac{1}{\tau}\frac{\partial \hat{a}}{\partial T} + \hat{a}\right) + b(\zeta, T).$$

Here $b(\zeta, T)$ is as yet arbitrary. At $O(\epsilon^3)$ the mean fields enter:

$$(16a) \qquad \frac{\partial \bar{\theta}_0}{\partial T} + \frac{1}{2}\frac{\partial}{\partial \zeta}(\hat{a}^2 \sin^2 \zeta) = \frac{\partial^2 \bar{\theta}_0}{\partial \zeta^2}$$

$$(16b) \qquad \frac{\partial \bar{\phi}_0}{\partial T} + \frac{1}{2\tau}\frac{\partial}{\partial \zeta}(\hat{a}^2 \sin^2 \zeta) = \tau \frac{\partial^2 \bar{\phi}_0}{\partial \zeta^2}.$$

These equations have the solution

$$(17) \qquad \bar{\theta}_0 = -d(T) \sin 2\zeta, \quad \bar{\phi}_0 = -c(T) \sin 2\zeta,$$

where

$$(18a) \qquad c_T - \frac{1}{2\tau}\hat{a}^2 = -4\tau c$$

$$(18b) \qquad d_T - \frac{1}{2}\hat{a}^2 = -4d.$$

Finally, at $O(\epsilon^4)$ the solvability condition for the fluctuating terms yields an expression for $b(\zeta, T)$ and the desired evolution equation for $\hat{a}$:

$$(19)$$
$$\hat{a}_{TT}\left[\frac{R_{S,CT}}{\tau^3} - R_{T,CT}\right] + \hat{a}_T\left[2\left(\frac{R_{S,CT}}{\tau^2} - R_{T,CT}\right) + \frac{1}{\sigma} + R_{T,2} - \frac{R_{S,2}}{\tau^2}\right]$$
$$+ \hat{a}\left[R_{T,2} - \frac{R_{S,2}}{\tau} - R_{T,4} + \frac{R_{S,4}}{\tau}\right] + R_{T,CT}\hat{a}d - \frac{R_{S,CT}}{\tau}\hat{a}c = 0.$$

Note that this is a second order equation. This is because of the proximity to the codimension-two point (9). Note also that even with the restriction $R_{T,2} - R_{S,2}/\tau =$

3 (see eq. (14)) one is able to vary the coefficients of $\hat{a}$ and $\hat{a}_T$ independently to unfold the bifurcation. Suitably scaled the final equations become

(20a)
$$a'' + \mu a' + \lambda a - ac + \kappa ad = 0$$

(20b)
$$c' = -\tau c + \frac{1}{\tau}a^2$$

(20c)
$$d' = -d + a^2,$$

where $\kappa \equiv (\sigma + \tau)/\tau(1 + \sigma)$ and the prime denotes differentiation with respect to T.

Equations (20) provide an asymptotic description of the p.d.e.'s in the sense that higher order terms vanish in the limit $\epsilon \downarrow 0$. It is possible to derive an even simpler system by taking in addition the asymptotic limit $\tau \downarrow 0$. For most fluids this is in fact a good approximation. If we let

(21a)
$$a = \tau^2 \tilde{a}, \quad c = \tau^2 \tilde{c}, \quad d = \tau^4 \tilde{d},$$

(21b)
$$\frac{\partial}{\partial T} = \tau \frac{\partial}{\partial \tilde{T}}, \quad \mu = -\tau \tilde{\mu}, \quad \lambda = \tau^2 \tilde{\lambda},$$

we find that, as $\tau \downarrow 0$, the mode d decouples, and we are left with the system (dropping tildes):

(22a)
$$a_{TT} - \mu a_T + \lambda a = ac$$

(22b)
$$c_T + c = a^2.$$

We call this third order system the *canonical* system. It is exact in the limit $\varepsilon \downarrow 0, \tau \downarrow 0$ and is the smallest possible system that can describe Shil'nikov dynamics. Its properties are summarized in Figure 3. The figure shows in the (μ, λ) plane the lines $\delta = \frac{1}{2}, 1$ as well as the line where the nontrivial fixed points have equal negative eigenvalues, all computed analytically. The locus of the heteroclinic connection was determined numerically and crosses the line $\delta = 1$ at $(\mu, \lambda) \simeq (0.15, 0.54)$. Consequently we expect the Shil'nikov dynamics to be present in region IV, and this is in accord with numerical integration of the system.

The canonical system can also be derived from the Galerkin expansion (11). These equations become exact at small amplitudes and in particular reduce to the Takens-Bogdanov normal form (10) in the neighborhood of the codimension-two point (9). For small τ these equations also simplify. Let $\hat{t} = \tau t$ be a slow time and

(23)
$$r = 1 + \mu \tau, \quad s = \tau^2 \nu, \quad a = \tau \hat{a}, \quad b = \tau \hat{b}, \quad c = \tau^2 \hat{c}.$$

Then $\hat{b} = \hat{a} - \tau \hat{a}' + O(\tau^2)$, and equations (11) reduce to [15]

(24a)
$$a' = ra - sd + O(\tau)$$

(24b)
$$d' = -d + a(1 - e) + O(\tau)$$

(24c)
$$e' = -\varpi e + \varpi ad + O(\tau),$$

where

$$r = \frac{\sigma\mu}{1+\sigma}, \quad s = \frac{\sigma\nu}{1+\sigma},$$

a replaces $\hat{a}$ and the prime denotes differentiation with respect to $\hat{t}$. These equations can be transformed into the Lorenz equations

(25a) $$x' = \tilde{\sigma}(y - x)$$

(25b) $$y' = \tilde{r}x - y - xz$$

(25c) $$z' = -\varpi z + xy,$$

where $\tilde{\sigma} = -r < 0$, and $\tilde{r} = s/r$.

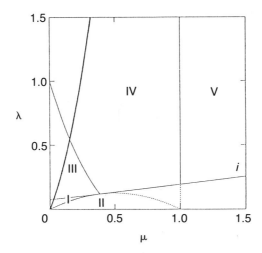

Figure 3. Eigenvalues at the non-stable fixed points of (22). Eigenvalues are real in regions I and II and complex elsewhere. The heavy curve shows the location of the heteroclinic bifurcation. Stable chaos is expected in region IV. (After Proctor & Weiss [21].)

When $\tilde{\sigma} < 0$ these equations behave quite differently from the régime described by Sparrow [23], and in particular contain Shil'nikov dynamics [4]. However, like equations (11), equations (24) or (25) are truncations and not rational approximations to the partial differential equations. Only in the limit $\varpi \downarrow 0$, obtained by defining the superslow time $t' = \varpi\hat{t}$ and setting

(26) $$r = 1 + \mu\varpi, \quad s - r = \lambda\varpi^2, \quad a = \varpi\tilde{a}, \quad d = \varpi b, \quad e = \varpi^2 c,$$

does one recover the canonical system (22) and hence obtains an asymptotic approximation to the p.d.e.'s. The third order system (24) does, however, provide a useful *model* of the p.d.e.'s when $\varpi = 0(1)$. Note that this argument shows that the limits $\varpi \downarrow 0$ and $\tau \downarrow 0$ commute.

Equation (26) shows that the canonical system is valid in an $O(\varpi^2)$ neighborhood of the line $r = s$, i.e., the locus of the pitchfork bifurcation from the trivial solution. Hence the Hopf bifurcation from the trivial solution occurring along $r = 1, s > 1$, is not described by the canonical system unless $s - 1 = O(\varpi^2)$. To describe what happens for fixed $s - 1 = O(\varpi)$ as r is varied it is necessary to use a different scaling. In contrast to (26) let $\tilde{t} = \varpi^{1/2}\hat{t}$ and

$$(27) \qquad r = 1 + \mu\varpi, \quad s = 1 + \nu\varpi, \quad a = \varpi^{1/2}\tilde{a}, \quad d = \varpi^{1/2}\tilde{d}, \quad e = \varpi\tilde{e}.$$

Dropping the tildes one now finds

$$(28a) \qquad a_{tt} + (\nu - \mu - e)a = O(\varpi^{1/2})$$

$$(28b) \qquad e_t = \varpi^{1/2}(-e + a^2) + O(\varpi).$$

Hence

$$(29) \qquad a = A\sin\Omega t, \quad \Omega^2 = \nu - \mu - e,$$

with the amplitude A and frequency Ω evolving on the time scale $T = \varpi^{1/2}t = \varpi\hat{t}$ according to

$$(30a) \qquad (A^2)_T = (\mu - \frac{1}{2})A^2 + (\nu - \mu)(A^2/2\Omega^2) - (3A^4/8\Omega^4)$$

$$(30b) \qquad (\Omega^2)_T = \nu - \mu - \Omega^2 - (A^2/2\Omega^2).$$

(see [15]). The periodic orbits are therefore given by

$$(31) \qquad A^2 = \frac{8\mu(\nu - \mu)}{1 + 4\mu}, \quad \Omega^2 = \frac{\nu - \mu}{1 + 4\mu};$$

they appear in the primary Hopf bifurcation at $\mu = 0$ and persist until $\nu = \mu$ (i.e., $s - r = O(\varpi^2)$) where their frequency (and amplitude) vanishes. This description thus captures the behavior between the Hopf bifurcation and the Shil'nikov dynamics described by equation (22).

It is possible to go the other way as well. Consider the canonical system in the limit $\lambda = (\nu - \mu)/\varpi \to \infty$. Let $t' = t/\lambda^{1/2}$ be the slow time and

$$(32) \qquad a = \lambda^{1/2}a', \quad c = \lambda c'.$$

After dropping primes equations (22) become

$$(33a) \qquad a_{tt} + a(1 - e) = \lambda^{-1/2}\mu a_t$$

$$(33b) \qquad e_t = \lambda^{-1/2}(-e + a^2).$$

Hence both $E \equiv a_t^2 + a^2(1 - e)$ and e vary slowly. With the superslow time $T = t/\lambda^{1/2} = t/\lambda$ we obtain the following averaged equations:

$$(34a) \qquad E_T = 2\mu\langle a_t^2\rangle - \langle a^4\rangle + e\langle a^2\rangle$$

$$(34b) \qquad e_T = -e + \langle a^2\rangle.$$

Hence

(35)
$$a = A \sin \omega t, \quad \omega^2 = 1 - e,$$

where

(36a)
$$E_T = \mu E - (3E^2/8\omega^4) + (1 - \omega^2)(E/2\omega^2)$$
(36b)
$$(\omega^2)_T = 1 - \omega^2 - (E/2\omega^2).$$

The limit cycle amplitude and frequency therefore satisfy

(37)
$$A^2 = \frac{8\mu}{1 + 4\mu}, \quad \omega^2 = \frac{1}{1 + 4\mu}.$$

It can now be checked that (31) and (37) agree if they are written in unscaled variables. Thus the outer limit of (22) matches onto the solution (29)-(31). We have therefore a complete description of the evolution of the oscillations with r for $s - 1 = O(\varpi)$. Figure 4 shows the variation of the period P with $r - 1$ for the two solutions, as given by (31) and (37), with $\varpi = 0.01$, and $s - 1 = 0.002$. The mismatch is $O(\varpi^{1/2})$, as expected.

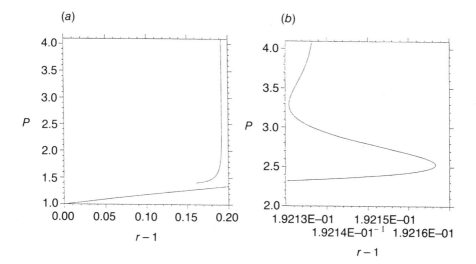

Figure 4. Period P as a function of $(r-1)$ for $\omega = 0.01, s-1 = 0.002$. (a) Inner and outer expansions given by (31) and (37). (b) Detail of behavior showing wiggly approach to heteroclinicity. (From Knobloch et al [15].)

III. Discussion. The analysis described above has been confined to oscillations in the form of standing waves that have the point symmetry (6). If this symmetry is relaxed one finds that typically the symmetric standing waves (hereafter SW) lose stability with increasing R_T to asymmetric SW. These secondary instabilities have been studied by Moore et al [19] for the p.d.e.'s and for a truncated Galerkin expansion describing magnetoconvection by Nagata et al [20]. Changing the boundary conditions from (5f) to the periodic boundary conditions

$$(38) \qquad \psi(x + 2\Lambda) = \psi(x), \quad \theta(x + 2\Lambda) = \theta(x), \quad \phi(x + 2\Lambda) = \phi(x)$$

has more dramatic consequences. In particular it introduces the symmetry group $O(2)$ of rotations and reflections of a circle into the problem. Since the primary instability breaks this symmetry the multiplicity of the pure imaginary eigenvalues at $R_T^{(o)}$ is doubled. As a consequence *two* branches of nontrivial oscillations bifurcate simultaneously from the trivial solution. These are the SW already described together with a branch of spatially periodic travelling waves (hereafter TW). In addition the SW, like the steady states SS, are no longer isolated; instead there is a circle of both, obtained by applying translations (mod 2Λ) to any one solution. With the boundary conditions (5e) at top and bottom the SW are almost always unstable to travelling wave disturbances and evolve into either left- or right-travelling waves [3,10]. With increasing R_T these waves also undergo secondary bifurcations. Near the Takens-Bogdanov bifurcation the TW must lose stability at a secondary Hopf bifurcation to modulated travelling waves (hereafter MW) before terminating on the unstable SS branch [2]. These MW terminate in a homoclinic bifurcation by colliding with the TW 1-torus. This result holds for the p.d.e.'s since it is derived from the normal form for the Takens-Bogdanov bifurcation with $O(2)$ symmetry. Further away from this bifurcation a Galerkin truncation using 15 (real) modes carried out for binary fluid convection shows that the MW terminates by becoming heteroclinic to the circles of SS and SW solutions [11]. Under certain conditions the MW 2-torus may undergo a cascade of torus-doubling bifurcations [12] leading to chaotic travelling waves. Whether this scenario provides an explanation for the chaotic TW described by Deane et al [3] remains to be seen.

Three remarks are in order. In the presence of the $O(2)$ symmetry the bifurcations from SW that break the reflection (6) lead to drifting SW (i.e. to MW). The Galerkin expansions are not as good at describing the TW as they are for SW. This is because the reflection in a vertical plane takes a left-travelling wave into a right-travelling one and hence does not force the existence of a solution with reflection symmetry. This is in contrast to SW or SS for which a reflection-symmetric solution always exists. Consequently, Galerkin truncations that are suitable for SW or SS impose an unphysical symmetry on the TW and hence, for a given truncation, will describe the TW less accurately [9]. This difficulty is easily overcome by enlarging the space of basis functions to include both $\sin n\pi x/\Lambda$ and $\cos n\pi x/\Lambda$. An additional difficulty arises with the boundary conditions (5e) at top and bottom.

With these the analogue of the truncation (11) takes the form

(39a)
$$\dot{a} = \sigma[-a + r_T b - r_S d]$$

(39b)
$$\dot{b} = -b + a(1 - c)$$

(39c)
$$\dot{c} = \varpi[-c + \frac{1}{2}(a\bar{b} + \bar{a}b)]$$

(39d)
$$\dot{d} = -\tau d + a(1 - e)$$

(39c)
$$\dot{e} = \varpi[-\tau e + \frac{1}{2}(a\bar{d} + \bar{a}d)],$$

where a, b, d are complex and c, e are real. Within this system the TW take the form

(40)
$$(a, b, c, d, e) = (a_0 e^{i\omega t}, b_0 e^{i\omega t}, c_0, d_0 e^{i\omega t}, e_0),$$

but exist at $r_T = r_T^{(0)}$ only. This degeneracy can be anticipated from the small amplitude theory [10] and implies that a larger number of modes is required in order that the TW branch be nondegenerate. Note finally that the limit $\tau \downarrow 0$ of the system (39) yields a system analogous to (24) but one in which the TW remain degenerate. Neither this system nor the complex version of (22) serve as useful models of $O(2)$-equivariant dynamics.

The procedure we have advocated for constructing Galerkin truncations utilizes all the modes that are generated to a particular order in perturbation theory around primary bifurcation points (e.g. [1,11]). The resulting truncation becomes *exact* at small amplitudes. The systems (11) and (39) provide simple examples. It is likely that the construction of approximate inertial manifolds will allow us to construct excellent (non-flat) Galerkin truncations of significantly lower order than those using the linear eigenfunctions. A particularly striking application of this technique can be found in Jolly et al [8] where a third order system (with six slaved modes) reproduces accurately the dynamics of the Kuramoto-Sivashinsky equation over a significant parameter range, including details of Shil'nikov dynamics. As mentioned above, care must be taken in any procedure not to force symmetries on solutions that do not possess them.

Acknowledgement. The work of E.K. was supported by NSF/DARPA under grant DMS-8814702.

REFERENCES

[1]   L.N. DaCosta, E. Knobloch and N.O. Weiss, *Oscillations in double-diffusive convection*, J. Fluid Mech. 109 (1981), pp. 25–43.

[2]   G. Dangelmayr and E. Knobloch, *The Takens-Bogdanov bifurcation with $O(2)$-symmetry*, Phil. Trans. Roy. Soc. London A 322 (1987), pp. 243–279.

[3]   A.E. Deane, E. Knobloch and J. Toomre, *Traveling waves and chaos in thermosolutal convection*, Phys. Rev. A 36 (1987), pp. 2862–2869.

[4]   J.N. Elgin and J.B. Molina Garza, *On the traveling wave solutions of the Maxwell-Bloch equations*, in Structure, coherence and chaos in dynamical systems, ed. P. Christiansen and R.D. Parmentier, Manchester Univ. Press (1988), pp. 553–562.

[5]   P. GLENDINNING AND C. SPARROW, *Local and global behavior near homoclinic orbits*, J. Stat. Phys. 35 (1984), pp. 645–696.

[6]   J. GUCKENHEIMER AND E. KNOBLOCH, *Nonlinear convection in a rotating layer: amplitude expansions and normal forms*, Geophys. Astrophys. Fluid Dyn. 23 (1983), pp. 247–272.

[7]   H.E. HUPPERT AND D.R. MOORE, *Nonlinear double-diffusive convection*, J. Fluid Mech. 78 (1976), pp. 821–854.

[8]   M.S. JOLLY, I.G. KEVREKIDIS AND E.S. TITI, *Approximate inertial manifolds for the Kuramoto-Sivashinsky equation: Analysis and computation*, Physica D 44 (1990), pp. 38–60.

[9]   E. KNOBLOCH, A.E. DEANE AND J. TOOMRE, *A model of double-diffusive convection with periodic boundary conditions*, Contemp. Math. 99 (1989), pp. 339–349.

[10] E. KNOBLOCH, A.E. DEANE, J. TOOMRE AND D.R. MOORE, *Doubly diffusive waves*, Contemp. Math. 56 (1986), pp. 203–216.

[11] E. KNOBLOCH AND D.R. MOORE, *A minimal model of binary fluid convection*, Phys. Rev. A 42 (1990), pp. 4693–4709.

[12] E. KNOBLOCH AND D.R. MOORE, *Chaotic travelling wave convection*, European J. Mech. B, in press.

[13] E. KNOBLOCH, D.R. MOORE, J. TOOMRE AND N.O. WEISS, *Transitions to chaos in two-dimen-sional double-diffusive convection*, J. Fluid Mech. 166 (1986), pp. 409–448.

[14] E. KNOBLOCH AND M.R.E. PROCTOR, *Nonlinear periodic convection in double-diffusive systems*, J. Fluid Mech., 108 (1981), pp. 291–316.

[15] E. KNOBLOCH, M.R.E. PROCTOR AND N.O. WEISS, *Heteroclinic bifurcations in a simple model of double-diffusive convection*, J. Fluid Mech., submitted.

[16] E. KNOBLOCH AND N.O. WEISS, *Bifurcations in a model of magnetoconvection*, Physica D 9 (1983), pp. 379–407.

[17] D.R. MOORE, J. TOOMRE, E. KNOBLOCH AND N.O. WEISS, *Period-doubling and chaos in partial differential equations for thermosolutal convection*, Nature 303 (1983), pp. 663–667.

[18] D.R. MOORE, N.O. WEISS AND J. WILKINS, *The reliability of numerical experiments: transitions to chaos in thermosolutal convection*, Nonlinearity 3 (1990), pp. 977–1014.

[19] D.R. MOORE, N.O. WEISS AND J. WILKINS, *Symmetry-breaking in thermosolutal convection*, Phys. Lett. A 147 (1990), pp. 209–214.

[20] M. NAGATA, M.R.E. PROCTOR AND N.O. WEISS, *Transition to asymmetry in magnetoconvection*, Geophys. Astrophys. Fluid Dyn. 41 (1990), pp. 211–241.

[21] M.R.E. PROCTOR AND N.O. WEISS, *Normal forms and chaos in thermosolutal convection*, Nonlinearity 3 (1990), pp. 619–637.

[22] L.P. SHIL'NIKOV, *A case of the existence of a countable number of periodic motions*, Sov. Math. Dokl. 6 (1965), pp. 163–166.

[23] C.T. SPARROW, *The Lorenz equations: Bifurcations, Chaos and Strange Attractors*, Springer-Verlag (1982).

DYNAMICAL STOCHASTIC MODELING OF TURBULENCE

ROBERT H. KRAICHNAN†

Abstract. Fully developed Navier-Stokes turbulence has proved resistant to methods of contemporary nonlinear dynamics that have been successful for small systems. An irreducibly large number of modes is excited in high-Reynolds-number turbulence. A set of tools that permits some progress is dynamical stochastic modeling, by which I mean the exact solution of model dynamical systems that have some relation to the true dynamics. The fact that a system is being solved exactly ensures some important consistency properties. Dynamical stochastic models whose tractability comes from intrinsic randomicity have given qualitatively and, in some cases, quantitatively good approximations to major features of fully developed turbulence: sensitivity to initial conditions, eddy viscosity, spectral energy cascade, and vorticity intensification. The present brief summary outlines three kinds of stochastic models: systems with random coupling coefficients, decimation models, and a new kind of stochastic model based on nonlinear mapping of Gaussian fields into dynamically evolving non-Gaussian fields.

Key words. turbulence, stochastic models

1. Introduction. Three-dimensional high-Reynolds-number Navier-Stokes (NS) turbulence embodies a complex set of phenomena. There is an interplay of randomness and order: well-formed, but plastic, structures, such as ropes of intense vorticity, exist within and interact with an ocean of more random-appearing excitation. The randomness is linked to an extreme sensitivity to perturbations. Nevertheless, key statistics are robustly stable against perturbation. It is a puzzle that such complicated-seeming dynamical behavior yields as much as it does to some crude ideas, notably eddy viscosity and step-wise cascade of energy.

The complexities of NS turbulence are associated with three related facts: the nonlinearity of the equations of motion; the simultaneous excitation of many degrees of freedom; the strong departure of hydrodynamic modes from absolute statistical-mechanical equilibrium. In this situation, it is unimportant whether the attractor is strange or merely complicated. The difference is undetectable and, moreover, the system shows most of its interesting features long before it has time to explore the attractor.

Much attention has focused on the nonlinear transfer of kinetic energy among wavevector components of the velocity field. This transfer arises from the shearing distortion of the velocity field by itself. A basic difficulty in systematic treatment of energy transfer at high wavenumbers (small spatial scales) is that the dominant terms in the NS equation for high-wavenumber components describe not distortion but simply the advection (carrying about) of small-scale structures by the large-scale velocity field. This advection usually is not associated with significant energy transfer. It is essential that theory distinguish consistently between the two effects.[14]

Turbulence theory needs to be able to handle finite Reynolds numbers! It is not adequate to predict only spectrum exponents of inertial ranges at infinite Reynolds

---

† 303 Potrillo Drive, Los Alamos, New Mexico 87544, U.S.A.

number. Many characteristic phenomena of turbulence, including intermittency of velocity derivatives, are present already at modest Reynolds numbers. Laboratory and computer experiments always are at finite Reynolds number, and there is a need to be able to make direct comparisons between theory and experiments, without guessing what asymptotic power laws the experiments seem to hint.

The models to be discussed in this paper are attempts to proceed systematically from the equations of motion to description and understanding of important physical processes in turbulent flows. They have had some success in showing how eddy viscosity, energy cascade, and other dynamical processes can arise from the equations of motion. In addition, the mapping models of Sec. 4 suggest mechanisms for the intermittency of small scales, a universal property of turbulent motion.

The guiding idea of dynamical stochastic modeling is to construct an exactly soluble dynamical system that exhibits some built-in invariance, conservation and realizability properties. The resulting statistics may be inaccurate or wrong, but the models cannot blow up or have grossly unphysical properties like negative energy spectra. The stochastic models to be reviewed in Sec. 2 are based on the inter-coupling of a collection of similar systems. This artifice creates an effective small parameter that vanishes in the limit of infinite collection, whatever the value of the Reynolds number. Consequently, the models can be solved exactly by perturbation methods despite the fact that such methods are not rationally applicable to the original NS dynamics. In Sec. 3, these models are imbedded in a broader class of models based on statistical treatment of the dynamical effects of the totality of modes in a large system upon small subsets of modes (decimation models). Decimation models give the hope of convergent sets of approximations to the exact NS statistics. Sec. 4 is devoted to stochastic models based on a novel and wholly non-perturbative approach: nonlinear mapping of Gaussian stochastic processes into dynamically evolving processes that may be very far from Gaussian. The resultant models predict entire probability distribution functions (PDF's) rather than moments and thereby can be used to attack the problem of intermittency of turbulent motion.

The point of view of the present summary is unabashedly that of theoretical physics. No attempt is made to give mathematically rigorous treatments. More detailed reviews of much of the following material have appeared.[3,13,14,16,18,20]

2. Random-coupling models.   The incompressible NS equation may be written

$$(1) \qquad \left(\frac{\partial}{\partial t} - \nu \nabla^2 \right) \mathbf{u}(\mathbf{x}, t) = -\lambda \mathbf{u}(\mathbf{x}, t) \cdot \nabla \mathbf{u}(\mathbf{x}, t) - \nabla p(\mathbf{x}, t),$$

$$(2) \qquad \qquad \nabla \cdot \mathbf{u} = 0,$$

where $\mathbf{u}$ is the velocity field, ν is kinematic viscosity, λ is an ordering parameter (equal to unity), p is pressure, and the uniform fluid density is set equal to unity. The advection term $\mathbf{u} \cdot \nabla \mathbf{u}$ dynamically couples velocity-field moments of different orders, with the result that the entire initial statistical distribution affects the evolution of any given moment over a finite time. This is commonly called the closure problem of turbulence theory. The linear viscous term also presents a closure problem, but one that shows up only in formulations for PDF's and is invisible at the level of moment analysis: the viscous term couples single-point PDF's to three-point PDF's.

In order to predict moments, attempts have been made to apply a variety of standard statistical approximations to (1) and to treat the nonlinear term by perturbation methods. Such treatments are essentially uncontrolled at moderate and large Reynolds numbers because the nonlinear term, which causes departure of moments from Gaussian values, is not small and the perturbation series are not convergent.[14]

A basic stochastic model for (1) can be constructed by first invoking a collection of M identical, but uncoupled, flow systems all residing in the same physical space. The equations of motion are

$$(3) \qquad \left(\frac{\partial}{\partial t} - \nu\nabla^2\right)\mathbf{u}^n(\mathbf{x},t) = -\lambda\mathbf{u}^n(\mathbf{x},t)\cdot\nabla\mathbf{u}^n(\mathbf{x},t) - \nabla p^n(\mathbf{x},t),$$

$$(4) \qquad \nabla\cdot\mathbf{u}^n = 0,$$

where $n = 1, 2, ..., M$ labels the different systems. The statistics of the individual flow systems are Gaussian, independent and identical at $t = 0$. The key steps now are to replace (3) by model dynamical equations that couple the M systems in a stochastic fashion and then to take the limit $M \to \infty$. The model equations are:

$$(5) \qquad \left(\frac{\partial}{\partial t} - \nu\nabla^2\right)\mathbf{u}^n(\mathbf{x},t) = -\lambda M^{-1}\sum_{rs}\phi_{nrs}\mathbf{u}^r(\mathbf{x},t)\cdot\nabla\mathbf{u}^s(\mathbf{x},t) - \nabla p^n(\mathbf{x},t),$$

with (4) unchanged. Here ϕ_{nrs} is a constant coefficient assigned the value $+1$ or -1 at random, subject only to invariance under any permutation of its three indices. The factor M^{-1} serves to make the variances of the total advection terms in (3) and (4) equal at $t = 0$. The ϕ_{nrs} vary only with the indices nrs. When a statistical ensemble of realizations of the entire collection of systems is formed, ϕ_{nrs} has precisely the same value in every realization in that ensemble.

Eq. (5) has been called the random-coupling (RC) model. It can be formulated in a number of different flavors. The effective small parameter is $1/M$. If $M \to \infty$, then (a) each individual interaction of three systems, represented by the terms in (5) involving a given ϕ_{nrs} and all of its index-permutations, can be treated as an infinitesimal perturbation on the total nonlinear interaction; (b) the cumulants that measure statistical dependence induced among any finite subset of flow systems vanish in the limit. The result is a closed set of equations that involve two fundamental statistical quantities: The velocity covariance in any of the flow systems and the mean response of the velocity amplitude in any flow system to an infinitesimal force added to the right side of (5). These quantities are independent of the flow-system labeling index n. The exactness of the perturbation analysis is a consequence solely of the limit $M \to \infty$ and does not depend on the size of the Reynolds number.[3]

The final closed equations for velocity covariance and mean infinitesimal response functions have been called the direct-interaction approximation (DIA), because the departures from independent Gaussian statistics arise from the direct interaction of small subsets (triads) of flow systems acting within the sea of interaction among all systems. As formulated above, the RC model and the DIA are not restricted to either isotropy or homogeneity. However, the final, closed equations are most easily written for homogeneous, isotropic statistics within a large cyclic box, and it is this case that will be discussed here.

Some important properties of the final DIA equations can be inferred from (5) without further analysis. First, (5) evolves actual mode amplitudes so that the resulting velocity covariance automatically satisfies all realizability inequalities. Second, the symmetry imposed on the ϕ_{nrs} maintains conservation of kinetic energy and other quadratic constants of motion under the nonlinear interaction. Third, the symmetry of the ϕ_{nrs} further assures that the absolute equilibrium canonical ensembles associated with mode truncation of the Euler equation carry over to the model.[7] A fourth essential property is plausible just from the fact that (5) describes a large, nonlinearly coupled system, but it must be verified by analysis. This is that (5), like (1), exhibits sensitivity to small perturbation when the Reynolds number is large. A fifth property is associated with a major deficiency of the DIA. This is the break-up of coherence effects in the self-advection of the velocity field. In (5), the term expressing advection of $\mathbf{u}^n(\mathbf{x}, t)$ by itself is replaced by a nonlinear term that involves all the other systems r and s. The result cannot be interpreted as advection of $\mathbf{u}^n(\mathbf{x}, t)$ by a model velocity field. The consequences are discussed later in this Section.

The final DIA integro-differential equations for homogeneous turbulence can be expressed in a compact and illuminating fashion by a Langevin equation for spatial Fourier modes. Take cyclic boundary conditions on a box of side $L \to \infty$ and introduce Fourier amplitudes by

(6)
$$u_i^n(\mathbf{x}, t) = \sum_{\mathbf{k}} u_i^n(\mathbf{k}; t) \exp(i\mathbf{k}\cdot\mathbf{x}),$$

The n-independent modal intensity scalar $U(k; t, t')$ may be defined by

(7)
$$\tfrac{1}{2} P_{ij}(\mathbf{k}) U(k; t, t') = (L/2\pi)^3 \langle u_i(\mathbf{k}; t) u_j^*(\mathbf{k}; t') \rangle \quad (L \to \infty),$$

where $\langle \rangle$ denotes ensemble average and $P_{ij}(\mathbf{k}) = \delta_{ij} - k_i k_j / k^2$ is a solenoidal projection operator. U is related to the energy spectrum in three dimensions by

(8)
$$E(k, t) = 2\pi k^2 U(k; t, t).$$

Eq. (5) rewritten in the Fourier representation becomes

(9)
$$\left(\frac{\partial}{\partial t} + \nu k^2 \right) u_i^n(\mathbf{k}; t) = -i\lambda \sum_{rs} \phi_{nrs} P_{ijm}(\mathbf{k}) \sum_{\mathbf{p}} u_j^r(\mathbf{p}; t) u_m^s(\mathbf{k} - \mathbf{p}; t),$$

where
$$P_{ijm}(\mathbf{k}) = \tfrac{1}{2}[k_m P_{ij}(\mathbf{k}) + k_j P_{im}(\mathbf{k})].$$

The DIA now can be expressed by the Langevin equation,[11,13]

$$(10) \qquad \left(\frac{\partial}{\partial t} + \nu k^2\right) u_i(\mathbf{k}; t) + \int_0^t \eta(k; t, s) u_i(\mathbf{k}; s) ds = b_i(\mathbf{k}; t),$$

Here $b_i(k; t)$ is a random, zero-mean forcing term given by

$$(11) \qquad b_i(\mathbf{k}; t) = -i\lambda P_{ijm}(\mathbf{k}) \sum_p \xi_j(\mathbf{p}; t) \xi_m(\mathbf{k} - \mathbf{p}; t),$$

where $\xi_i(\mathbf{k}; t)$ is an isotropic, homogeneous Gaussian velocity field that satisfies

$$(12) \qquad \langle \xi_i(\mathbf{k}; t) \xi_j^*(\mathbf{k}; t') \rangle = \langle u_i(\mathbf{k}; t) u_j^*(\mathbf{k}; t') \rangle.$$

Thus $b_i(\mathbf{k}; t)$ has precisely the form of the entire nonlinear term in the wavevector form of the original NS equation. However, in distinction to the exact $\mathbf{u}(\mathbf{x}, t)$, the field $\boldsymbol{\xi}(\mathbf{x}, t)$ remains precisely Gaussian under the dynamics. The $\eta(k; t, s)$ term in (10) is a dynamical damping term that balances energy input from the forcing term $b_i(\mathbf{k}; t)$ and thereby maintains energy conservation by the nonlinear dynamics. It is given by

$$(13) \qquad \eta(k; t, s) = \pi k \lambda^2 \iint_{\triangle} pq \, dp \, dq \, b(k, p, q) G(p; t, s) U(q; t, s),$$

Here $G(k; t, t')$ describes the response of $u_i(\mathbf{k}; t)$ to an infinitesimal source term added to the right side of (10), $\iint_{\triangle}$ denotes integration over all wavenumbers p, q such that k, p, q can form a triangle, x, y, z are the cosines of the internal angles opposite sides k, p, q, and

$$(14) \qquad b(k, p, q) = (p/k)(xy + z^3).$$

Note that (10) is a linear dynamical equation in any single realization because $\eta(k; t, s)$ and $b_i(\mathbf{k}; t)$ are determined by ensemble averages and are unaffected by the value of $\mathbf{u}(\mathbf{k}; s)$ in any one realization.

The contribution to the $\eta(k; t, s)$ term in (10) from $p, q \gg k$ in (13) has precisely the form of a viscous damping: under the assumption that characteristic times of modes p and k are much smaller than those of modes k, this contribution is $\propto \nu(k, t) k^2 u_i(\mathbf{k}, t)$, where $\nu(k, t)$ is an effective eddy viscosity computed from (13). Thus the DIA gives analytical support to the concept of eddy viscosity. However, the following must be noted: first, $\nu(k, t)$ varies with k, in general; second, if the inequality $p, q \gg k$ is not satisfied, the associated contributions to the η term in (10) are nonlocal in time. Exact expressions for eddy-damping coefficients that describe the action of turbulence on weak external velocity gradients may be obtained without appeal to statistical approximations. The DIA results represent approximate reductions of these expressions and extend the eddy-damping formalism to strong field gradients.

The interpretation of the η term in (10) as an eddy-damping term fails completely for the contributions to (13) from $p \ll k$ or $q \ll k$. In this case, it can be shown that the associated contributions to the η term and b term in (10) nearly cancel each other. There is nearly no net contribution to energy flow in or out of modes k associated with these contributions. The physical meaning of this result is that the action of large spatial scales of motion on small spatial scales is primarily to carry the latter bodily, without significant distortion. It is the distortion of flow structures by straining that gives rise to net energy cascade. Even though large spatial scales (low wavenumbers) may dominate the energy of the flow, they usually contribute little to overall shear and strain.

The distinction between advection and distortion effects is clarified by examining the transformation properties of solutions of the NS equation in a cyclic box under a random Galilean transformation (RGT). The latter consists of adding to the initial velocity field, in each of an ensemble of solutions, a spatially uniform random velocity whose amplitude and direction are chosen independently in each realization. Each solution thereby remains a solution but suffers a spatial translation that increases uniformly with time. No energy transfer among Fourier modes is induced by the added uniform velocity because this velocity does not distort. Thus, energy transfer is invariant under RGT.

The cancellation of low-wavenumber contributions to the η and b terms noted above shows that, at the grossest level, the DIA and associated RC model distinguish between convection and distortion effects. Nevertheless the DIA energy transfer is not invariant under RGT because of a subtler failing:[8] RGT changes the contributions to η and b involving finite p and q and thereby changes the energy transfer among the triad of wavenumber modes k, p, q. This happens because of the nature of the functions G and U that are fundamental to DIA. As a function of $t - t'$, $U(k; t, t')$ is a measure of phase decorrelation of a Fourier component of the Eulerian velocity field. A RGT enhances this phase decorrelation because it sweeps the Eulerian velocity field past fixed points in space. There is a similar behavior of $G(k; t, t')$ under RGT. DIA correctly captures this effect of RGT on the decay of $U(k; t, t')$ and $G(k; t, t')$ with $t - t'$. However, the DIA approximation for energy transfer is a simple expression in U and G and, as a consequence, is decreased under RGT. This effect is spurious and violates invariance under RGT.

Invariance of energy transfer under RGT can be restored by modifying the DIA so that $U(k; t, t')$ and $G(k; t, t')$ are replaced by functions built from Lagrangian rather than Eulerian fields (Lagrangian-history DIA or LHDIA).[4,9,12] However the resulting approximation no longer is derivable from a stochastic model. Alternatively, a stochastic model that gives invariance of energy transfer under RGT can be constructed by the decimation techniques to be described in Sec. 3.

If one assumes in advance that interactions with much smaller wavenumbers should not affect energy transfer in or out of modes with a wavenumber k, then it is easy to construct RC models that are RGT-invariant. One can modify the coupling coefficient $P_{ijm}(\mathbf{k})$ in (9) so that sweeping effects are removed from the interaction of k, p, q (p or $q \ll k$) and only eddy-damping effects are retained.[8] The RC model and DIA approximation are then exactly RGT-invariant. Such a modification has also been implicit in renormalization-group turbulence approximations, which effectively discard sweeping of small scales by large scales at the outset.[16]

At low and moderate Reynolds numbers, the DIA equations for isotropic turbulence have given good qualitative and quantitative results for energy spectrum, energy transfer, and the dependence of $U(k; t, t')$ on $t - t'$.[16] At larger Reynolds numbers the predictions are flawed by the violation of RGT invariance. A consequence is that DIA gives an inertial-range spectrum of form $E(k) \propto k^{-3/2}$ at infinite Reynolds number,[5] while experiment suggests that the asymptotic spectrum is close to the Kolmogorov form $E(k) \sim C\epsilon^{2/3} k^{-5/3}$, where ϵ is the rate of energy dissipation by viscosity, per unit mass. The LHDIA approximation, which keeps invariance under RGT, yields the Kolmogorov form. In its simplest form, it gives both a value of the Kolmogorov constant C and a shape of $E(k)$ in the dissipation range of wavenumbers k that agree well with experiments at high Reynolds numbers.[9]

3. Decimation models. One might hope that closed approximations above DIA could be constructed by giving to the coupling coefficients ϕ_{nrs} in (5) or (9) values that somehow lie between the totally random symmetric values of the RC model and the value $\phi_{nrs} = M\delta_{nr}\delta_{ns}$ that gives back the original NS dynamics (3). Such models have not materialized. However, it is possible to imbed DIA in a set of systematically improved approximations by constructing stochastic models of a different kind, again starting from the collection of NS systems (3).

The collection of M systems may be described by a set of M statistically symmetrical collective fields that are orthogonal linear combinations of the $\mathbf{u}^n(\mathbf{x}, t)$. The models are then constructed by a several-stage process: First the collective fields are divided into an explicit set, with $S \leq M$ members and an implicit set containing the remaining $M - S$ collective fields. The explicit set of field amplitudes is to be followed in detail. Second, the nonlinear terms representing the action of the implicit set in the equations of motion of the explicit set are considered to be forcing terms that are known only statistically. Third, the statistics of these forcing terms are systematically constrained toward their true values by moment equations, or more general equations, that express the statistical symmetry between the explicit amplitudes and the implicit amplitudes. This permits a bootstrap procedure in which the dynamics of the explicit collective fields is followed as they interact with each other amid the sea of interactions with the implicit fields.

The collective fields may be defined by:[6]

$$(15) \quad \mathbf{u}^\alpha(\mathbf{x}, t) = M^{-1/2} \sum_n e^{i2\pi\alpha n/M} \mathbf{u}^n(\mathbf{x}, t) \quad [\alpha = 0, \pm 1, \pm 2, ..., \pm(M-1)/2],$$

where Latin superiors label individual fields in the collection (as before), Greek superiors label the collective fields, and the collection size M is taken odd. The orthogonality identities

$$(16) \quad \sum_\alpha e^{i2\pi\alpha(n-m)/M} = M\delta_{nm}, \quad \sum_n e^{i2\pi(\alpha-\beta)n/M} = M\delta_{\alpha\beta},$$

make (15) equivalent to

$$(17) \qquad \mathbf{u}^n(\mathbf{x}, t) = M^{-1/2} \sum_\alpha e^{-i2\pi\alpha n/M} \mathbf{u}^\alpha(\mathbf{x}, t).$$

With the aid of (15)–(17), Eqs. (3) and (4) for the collection of statistically identical NS flow systems can be transformed to the following equations for the collective fields:

$$(18) \qquad \left(\frac{\partial}{\partial t} - \nu\nabla^2\right) \mathbf{u}^\alpha(\mathbf{x}, t) = -\lambda M^{-1/2} \sum_{\beta+\gamma=\alpha} \mathbf{u}^\beta(\mathbf{x}, t) \cdot \nabla \mathbf{u}^\gamma(\mathbf{x}, t) - \nabla p^\alpha(\mathbf{x}, t),$$

$$(19) \qquad\qquad\qquad \nabla \cdot \mathbf{u}^\alpha = 0.$$

The exact NS dynamics are unaltered by this transformation. The description of the M statistically identical flow systems by the collective fields is formally analogous to the description of a spatially homogeneous dynamical system, with finite coherence lengths, by spatial Fourier modes in a cyclic box. Like the Fourier modes of a homogeneous system, the collective fields yield nonvanishing moments only if the labeling superscripts add to zero. For example,

$$(20) \qquad \langle \mathbf{u}^\alpha(\mathbf{x}, t)\mathbf{u}^\beta(\mathbf{x}', t') \rangle = \delta_{\alpha+\beta} \langle \mathbf{u}^n(\mathbf{x}, t)\mathbf{u}^n(\mathbf{x}', t') \rangle,$$

where $\delta_\mu \equiv 1$ if $\mu = 0$, $\delta_\mu \equiv 0$ otherwise, and n denotes any one system from the collection of statistically identical systems. Eq. (20) illustrates that statistical properties of individual collective fields are independent of α. In general, the ensemble average of any product of collective field amplitudes with a set of labeling superscripts α, β, ... that adds to zero depends only on how this set decomposes into zero-sum subsets; the value is otherwise independent of the particular values of α, β, ...

RC models may be constructed by altering (18); if fact, this was the original method of construction.[6] To do this, factors $\phi_{\alpha\beta\gamma}$ are inserted after the summation sign in (18). They take the values $+1$ and -1 at random subject to the symmetry requirements

$$(21) \qquad \phi_{\alpha\beta\gamma} = \phi_{\alpha\gamma\beta} = \phi_{\beta\alpha,-\gamma} = \phi_{-\alpha,-\beta,-\gamma},$$

which maintain energy conservation and the reality of all the $\mathbf{u}^n(\mathbf{x}, t)$. The RC model so constructed is not identical with that formed directly from (3), but in the limit $M \to \infty$ both models yield the same statistics for each $\mathbf{u}^n(\mathbf{x}, t)$ and, in particular, the same DIA equations.

The present Section is concerned with the construction of another kind of model: the decimation (DEC) models which exploit, in a different way, the statistical symmetry of the collection of systems. The DEC models are constructed by dividing the total set of M collective fields into an explicit, or sample, subset with S members and an implicit subset with $M - S$ members.[13,16,17,22] If α is in the explicit subset, then $-\alpha$ is also. The equations of motion for the explicit subset are written as

$$\left(\frac{\partial}{\partial t} - \nu\nabla^2\right)\mathbf{u}^\alpha(\mathbf{x}, t) =$$

$$(22) \qquad -\lambda M^{-1/2} \sum_{\beta+\gamma=\alpha}^{S} \mathbf{u}^\beta(\mathbf{x}, t)\cdot\nabla\mathbf{u}^\gamma(\mathbf{x}, t) - \nabla p^\alpha(\mathbf{x}, t) + \mathbf{q}^\alpha(\mathbf{x}, t) \quad (\alpha \in S),$$

where the S-sum is only over β, γ pairs that lie in the explicit subset and $\mathbf{q}^\alpha(\mathbf{x}, t)$ then represents all the terms in the sum in (22) such that β and/or γ are outside the explicit set. So far the exact dynamics are undisturbed. The models are now constructed by considering $\mathbf{q}^\alpha(\mathbf{x}, t)$ to be a stochastic field constrained only by statistical relations that express the underlying statistical symmetry among the collective fields. In this way, moments, or other statistics, of $\mathbf{q}^\alpha(\mathbf{x}, t)$ are expressed in terms of statistics of the explicit fields alone. The statistical determination replaces determination of $\mathbf{q}^\alpha(\mathbf{x}, t)$ from the actual equations of motion for the implicit collective fields.

The model construction is best illustrated by an example. Let the explicit set consist of the single member $\alpha = 0$. Then (22) becomes simply

$$\left(\frac{\partial}{\partial t} - \nu\nabla^2\right)\mathbf{u}^0(\mathbf{x}, t) =$$

$$(23) \qquad -\lambda M^{-1/2} \sum_\beta \mathbf{u}^\beta(\mathbf{x}, t)\cdot\nabla\mathbf{u}^{-\beta}(\mathbf{x}, t) - \nabla p^0(\mathbf{x}, t) + \mathbf{q}^0(\mathbf{x}, t).$$

The statistical symmetry among the collective fields implies

$$(24) \qquad \langle\mathbf{q}^0(\mathbf{x}, t)\mathbf{u}^0(\mathbf{x}', t')\rangle = (M-1)\langle\mathbf{u}^0(\mathbf{x}', t')\mathbf{u}^0(\mathbf{x}, t)\cdot\nabla\mathbf{u}^0(\mathbf{x}, t)\rangle.$$

This is because the average on the left of (24) consists of $M-1$ terms, each equal to the average on the right side according to the statistical symmetry.

The constraint (24) is of fundamental importance because for $t = t'$ it constrains $\mathbf{q}^0$ so that the total nonlinear interaction [$\mathbf{q}(\mathbf{x}, t)$ plus the explicit sum in (23)] conserves mean energy. This single moment constraint is insufficient to fully determine the statistics of $\mathbf{q}^0$. But the determination can be made unambiguous by requiring that $\mathbf{q}^0(\mathbf{x}, t)$ have the minimum variance consistent with (23) and (24). In the limit $M \to \infty$, it can be shown that this choice yields precisely the DIA, in the form of the Langevin equation (10). The amplitude $u_i(\mathbf{k}; t)$ in (10) now represents the Fourier transform of $\mathbf{u}^0(\mathbf{x}, t)$ while the transform of $\mathbf{q}^0(\mathbf{x}, t)$ (with pressure contribution incorporated) is

$$(25) \qquad q_i^0(\mathbf{k}; t) = b_i(\mathbf{k}; t) - \int_0^t \eta(k; t, s)u_i(\mathbf{k}; s)ds.$$

Thus (10) can be interpreted as the result of realizing $\mathbf{q}^0(\mathbf{x}, t)$ in the limit $M \to \infty$ with the least variance permitted by the symmetry constraint (24).

The advantage of constructing DIA *via* the DEC model is that systematically improved models can be constructed by adjoining higher statistical symmetry constraints to (24). These constraints are all obeyed exactly in the actual NS dynamics. If all possible symmetry constraints are imposed on $q^0(\mathbf{x}, t)$, the resulting model reproduces all NS statistics. A model logically one step above DIA is obtained by adjoining to (24) the constraints that express $\langle q^0(\mathbf{x}, t)\mathbf{u}(\mathbf{x}', t')\mathbf{u}(\mathbf{x}'', t'')\rangle$ and $\langle q^0(\mathbf{x}, t)q^0(\mathbf{x}', t')\rangle$ in terms of moments of $\mathbf{u}^0(\cdot, \cdot)$ alone. It has been shown that these constraints are sufficient to enforce invariance of energy transfer under RGT.

In field-theoretic terminology, DIA corresponds to a lowest-order truncation of a line-renormalized expansion for triple moments of the velocity field. The higher DEC models do not appear to have any simple correspondence to line- or vertex-renormalized perturbation expansions. Since the imposed constraints are always ones obeyed in the exact NS dynamics, successively higher DEC models are expected to yield statistics that converge to exact NS statistics. This is in contrast to the divergence of truncations of renormalized perturbation expansions. There has been little exploration of higher DEC models, but convergence has been verified in an application to an idealized convection problem.

4. Mapping Models. The stochastic models so far described in this paper all are models of dynamics: the actual equations of motion are replaced by altered equations of motion chosen so that exact values of certain statistics can be obtained. Models of this kind, in particular those related to DIA, have been successful in predicting second- and third-order moments of the velocity field; they have had limited success in predicting departures of fourth-order moments from Gaussian values;[2,21] and they have totally failed to capture the intermittency of small scales.[2,10] There is little reason to believe that such models will be of value in approximating the shapes of PDF's of the amplitudes of velocity and its spatial derivatives.

A basic shortcoming of the RC model in handling PDF's can be seen immediately from the equation of motion (5). In the limit $M \to \infty$, necessary to get analytical results for the model, each field $\mathbf{u}^n(\mathbf{x}, t)$ is coupled to an infinite number of other fields. Arguments based on the central limit theorem then suggest that the statistics of each individual $\mathbf{u}^n(\mathbf{x}, t)$ are Gaussian in the limit. This can be verified. The joint PDF of all the $\mathbf{u}^n(\mathbf{x}, t)$ of course is not Gaussian. Thus, the model values for triple moments are nonzero, but they are obtained by summing over all the system labels, and they contain the random ϕ coefficients. There does not appear to be any simple way to construct, from the model, a non-empty approximation to PDF's associated with a single field – for example, the probability that the velocity field amplitude at a single space-time point has a given value. Related difficulties arise when low-order decimation models are asked to provide approximations to PDF's.

There is an alternative approach: model the stochastic velocity field itself in x space and apply the exact dynamical equations to the model field. Current work suggests that valid and workable approximations to PDF's can be constructed in this way and that this approach offers powerful, non-perturbative tools for handling strong turbulence. The stochastic models to be discussed now are constructed by nonlinear mappings that carry a multivariate Gaussian "reference field" into a field that can be wildly non-Gaussian. Closure approximations are obtained by exploiting the known statistics of the reference field.[1,15,16,19] The reference field need not be Gaussian; the basic requirement is that its statistics be fully known and accessible.

The moment closures discussed already in this paper are based on the fact of polynomial nonlinearity, in particular quadratic nonlinearity. But the mapping closures to be discussed now are substantially indifferent to the form of the nonlinearity; the latter need not be algebraic. For this reason, the mapping closures are well suited for application to compressible flow with real equations of state and to other problems that are daunting to moment closures.

The modeling by mapping will be illustrated here by application to a problem simpler than NS dynamics: the initial development of intermittency in Burgers turbulence. Statistical homogeneity is assumed, but the mapping method is also applicable to inhomogeneous flows in any geometry.

Burgers' equation may be written

$$(26) \qquad \mathcal{D}u/\mathcal{D}t = \nu u_{xx}, \qquad \mathcal{D}\xi/\mathcal{D}t = -\xi^2 + \nu\xi_{xx},$$

where $u(x,t)$ is velocity, $\mathcal{D}/\mathcal{D}t \equiv \partial/\partial t + u\partial/\partial x$ and $\xi \equiv u_x \equiv \partial u/\partial x$.

The dominant behavior under Burgers' equation is steepening of negative velocity gradients into viscosity-limited shocks. During this process an initially Gaussian field u retains a nearly Gaussian univariate distribution $P(u)$ while the univariate distribution $Q(\xi)$ of ξ becomes highly intermittent, even at low Reynolds number.

Analytical mapping closure starts with a multivariate-Gaussian *reference field*; for the Burgers example, the reference velocity field may be taken as $u_0(z)$. It is assumed that the actual field $u(x,t)$ and laboratory position x are related to u_0 and the reference coordinate z by

$$(27) \qquad u = X(u_0,t), \qquad dz/dx = J(u_0,\xi_0,t),$$

where X and J are ordinary, non-stochastic functions and the arguments u, u_0 and ξ_0 all are values measured at the same point in space (any point z). Thus

$$(28) \qquad \xi = \xi_0 J(u_0,\xi_0)\partial X/\partial u_0 \equiv Y(u_0,\xi_0).$$

where $\xi_0 \equiv \partial u_0/\partial z$. Eqs. (27) and (28) describe two kinds of nonlinear distortion that are characteristic of mapping closure: transformation of amplitudes (the function X) and change of measure (the Jacobian J of the coordinate transformation from z to x). Both are determined at each point in z space by local properties – in the present case, u_0 and ξ_0.

The analysis is simplified by the approximation that $P(u)$ is Gaussian and that u and ξ are statistically independent. Therefore it will be assumed that the single-point joint PDF of u and ξ for the model field has the form

$$(29) \qquad P(u,\xi) = P(u)Q(\xi),$$

(30) $$u = X(u_0, t) = r(t)u_0, \qquad \xi = r(t)\xi_0 J(\xi_0) \equiv Y(\xi_0, t).$$

Here $r(t)$ measures the decay of velocity amplitudes under viscosity. Energy conservation then yields

(31) $$dr/dt = -\nu r \langle \xi^2 \rangle / \langle u^2 \rangle.$$

The X transformation in (30) is now a trivial linear transformation. For applications in which X is nontrivial, see [16]. The analysis for Burgers' equation can be extended to the joint PDF $P(u, \xi)$, without the assumption of linear X.

It is easily shown that $Q(\xi)$, as defined by (29) and (30), exactly obeys the reduced Liouville equation

(32) $$\frac{\partial Q(\xi)}{\partial t} + \frac{\partial}{\partial \xi}\left(\left[\frac{D\xi}{Dt}\right]_{C:\xi} Q(\xi)\right) = \xi Q(\xi),$$

where $[\ ]_{C:\xi}$ denotes ensemble mean conditional only on a given value ξ and the divergence term on the right side expresses the difference of measure between Lagrangian and Eulerian coordinates. On the other hand, (27)–(30) give

(33) $$Q(\xi) = Q_0(\xi_0)\left(\frac{\partial Y}{\partial \xi_0}\right)^{-1}\frac{N}{J}.$$

where $Q_0(\xi_0)$ is the Gaussian PDF of the gradient of the reference field, the factor $1/J$ expresses the change of measure associated with squeezing or stretching of z to give x, and $N(t)$ normalizes $Q(\xi)$ to unity. Eq. (33) implies the reduced Liouville equation

(34) $$\frac{\partial Q(\xi)}{\partial t} + \frac{\partial}{\partial \xi}\left(\frac{\partial Y}{\partial t}Q(\xi)\right) = \alpha(\xi)Q(\xi),$$

where $\alpha(\xi) = \partial \ln(N/J)/\partial t$. The requirement that (34) give the same $Q(\xi)$ as (32) leads to an evolution equation for J:

$$\frac{\partial J}{\partial t} = -r\xi_0 J^2 - \frac{1}{r\xi_0 Q(\xi)}\int_{-\infty}^{\xi}[\alpha(\xi') - \xi']Q(\xi')d\xi'$$

(35) $$+\nu(r\xi_0)^{-1}[\xi_{xx}]_{C:\xi} + \nu J \langle (\xi_0 J)^2 \rangle / \langle u_0^2 \rangle,$$

It should be noted that the integrals of the right sides of (33) and (34) over ξ vanish, so that total probability is conserved. This follows for (33) from statistical homogeneity and for (34) from the definition of α.

To obtain closure, $[\xi_{xx}]_{C:\xi}$ must be obtained. The evaluation can be done exactly by performing chain differentiation of (30) and using the known statistics of the reference field – in particular, the fact that the amplitude and gradient of a homogeneous Gaussian field at a point are statistically independent.[1,16] The result is

(36) $$[\xi_{xx}]_{C:\xi} = -r\xi_0 k_d^2\left(J^3 + \frac{\xi_0}{3}\frac{\partial J^3}{\partial \xi_0}\right) + rC_2\left(\frac{\partial J^3}{\partial \xi_0} + \frac{\xi_0}{2}J\frac{\partial^2 J^2}{\partial \xi_0^2}\right).$$

with $C_2 = \langle(\partial\xi_0/\partial z)^2\rangle$, $k_d^2 = C_2/\langle\xi_0^2\rangle$.

The J^2 term in (35) comes from the $-\xi^2$ term in (26). The derivative terms on the right side of (36) come from exact treatment of ξ_{xx} under the space-varying distortion J. The integral term in (35) arises from the effect on $P(u,\xi)$ of the N/J factor in (33); it makes (35) an integro-differential equation that must be solved iteratively. The derivative and integral terms play an essential role in shaping $Q(\xi)$ near its maximum. The qualitative behavior of J at large negative ξ is controlled by the J^2 term in (35), which expresses the amplification of negative gradients, and the J^3 term in (36), which expresses viscous relaxation of gradients. The result is that $Q(\xi)$ exhibits a tail of form $|\xi|^{-1}\exp(-\text{const}|\xi|)$ for large negative ξ.

$P(u,\xi)$ is unchanged if the transformation $r(t)$, $J(\xi_0,t)$ is replaced by three successive measure-preserving operations on initial field-realizations: (a) an effective viscous relaxation of the reference field that changes $\langle\xi_0^2\rangle/\langle u_0^2\rangle$ by the factor $[N(t)]^2$ and $\langle u_0^2\rangle$ by a related factor $[r_0(t)]^2$; (b) a squeezing of the relaxed reference field by the factor $J_N(\xi_0,t) = J(\xi_0,t)/N(t)$; (c) a reduction of amplitudes by the factor $r(t)/r_0(t)$.

The $Q(\xi)$ obtained from solution of (29)–(31), (33)–(36) has been found to give excellent quantitative agreement with direct simulations of (26) in which departure from Gaussian shape is marked.[15] Nothing is adjustable in the closure.

Burgers and NS dynamics are very different under mapping closure as Reynolds number approaches infinity. The sharpening of shocks as $\nu \to 0$ implies that $K_\xi = \langle\xi^4\rangle/\langle\xi^2\rangle^2$ increases indefinitely with Reynolds number. This is reflected in the analytical form of the closure PDF: The prefactor $|\xi|^{-1}$ makes $Q(\xi)$ non-normalizable if the $|\xi| \to \infty$ form is extended to $\xi = 0$. Both closure solution and simulations give increasingly sharp peaks at the maximum of $Q(\xi)$ as Reynolds number increases.

Mapping closures evolve each realization in the initial Gaussian ensemble so that multi-point statistics, and spectra, may be computed over the evolved ensemble. In the Burgers case at large Reynolds number, sharpening of negative gradients into viscosity-limited shocks is exhibited by the model realizations, and the inertial-range spectrum exponent is $n = -2$.

Analytical mapping closures for NS dynamics must deal explicitly with the pressure term in (1). If the flow is compressible, the NS equation can be augmented by an equation of state, a continuity equation for fluid density, and a heat or entropy equation. The pressure is then a function of local quantities and generalization of the mapping closure techniques is essentially straightforward. In the case of incompressible NS flow, pressure is intrinsically nonlocal (the solution of a Poisson equation). There are several devices for dealing with this problem within the framework of locally-determined mapping transformations. One alternative is to eliminate the pressure entirely and deal with velocity vector-potentials that are time-dependent functions of reference fields and their low-order spatial derivatives. Another alternative is to introduce an approximate pressure that is a local function of velocity and low-order spatial derivatives of velocity. This leads to velocity fields that are approximately solenoidal. In any event, a prime challenge is the sharp difference in statistics between NS turbulence in two and three dimensions.

This work was supported by the National Science Foundation, Division of Atmospheric Sciences, under Grant ATM-8807861 to Robert H. Kraichnan, Inc. and by the Department of Energy under Contract W-7405-Eng-36 with the University of California and subcontracts thereunder between the University and Robert H. Kraichnan, Inc. During part of this work, the author served as consultant to the Theoretical Division, Los Alamos National Laboratory.

REFERENCES

[1] H. CHEN, S. CHEN AND R. H. KRAICHNAN, Phys. Rev. Lett. 63 (1989), p. 2657.

[2] H. CHEN, J. R. HERRING, R. M. KERR AND R. H. KRAICHNAN, Phys. Fluids A, 1 (1989), p. 1844.

[3] J. R. HERRING AND R. H. KRAICHNAN, in *Statistical Models and Turbulence*, edited by M. Rosenblatt and C .Van Atta, Springer-Verlag, New York, 1972.

[4] Y. KANEDA, J. Fluid Mech. 107 (1989), p. 131.

[5] R. H. KRAICHNAN, J. Fluid Mech. 5 (1959), p. 497.

[6] —, J. Math. Phys. 2 (1961), p. 124.

[7] —, Phys. Fluids 7 (1964), p. 1030.

[8] —, Phys. Fluids 7 (1964), p. 1723.

[9] —, Phys. Fluids 9 (1966), p. 1728.

[10] —, Phys. Fluids 10 (1967), p. 2081.

[11] —, J. Fluid Mech. 41 (1970), p. 189.

[12] —, J. Fluid Mech. 83 (1977), p. 349.

[13] —. in *Theoretical Approaches to Turbulence*, edited by D. L. Dwoyer, M. Y. Hussaini and R. G. Voight, Springer-Verlag, New York, 1985.

[14] — in *Current Trends in Turbulence Research*, edited by H. Branover, M. Mond and Y. Unger, American Inst. of Aeronautics, Washington, 1988, p. 198.

[15] —, Phys. Rev. Lett. 65 (1990), p. 575.

[16] — in *New Perspectives in Turbulence*, edited by L. Sirovich, Springer-Verlag, New York, 1990.

[17] — AND S. CHEN, Physica D 37 (1989), p. 160.

[18] D. C. LESLIE, *Developments in the Theory of Turbulence*, Oxford University Press, 1973.

[19] S. B. POPE, private communication.

[20] H. A. ROSE AND P. L. SULEM, J. Physique 39 (1978), p. 441.

[21] L. SHTILMAN AND W. POLIFKE, Physics of Fluids A, 1 (1989), p. 778.

[22] T. WILLIAMS, E. R. TRACY AND G. VAHALA, Phys. Rev. Lett. 59 (1987), p. 1922.

ON A NEW TYPE OF TURBULENCE
FOR INCOMPRESSIBLE MAGNETOHYDRODYNAMICS

V. P. MASLOV

1. By passing to the limit in a certain physical process we often obtain evolutionary partial differential equations in the space $R_+ \times R^n$. And we want to know how small perturbations in initial conditions effect the prosses of passing to the limit with respect to ε which is assumed to be a small parameter. If a change of initial data by a value $\mu\phi(x)$, $x \in R^n$, ($\phi(x)$ is an entire bounded function independent of μ and ε) leads in time $t < \delta$ to a change of the solution by a value of order 1 (μ and δ are arbitrary small independent of ε), then the incorrectness (ill-posedness) is so strong that it cannot be regularised. This means that the passing to the limit is incorrect itself. Such incorrectness will be called the burst incorrectness.

An important process of passing to the limit is the averaging over a small volume and a small interval of time $\int_{\delta_{x_0,t^0}} u_\varepsilon(x,t)\,dx\,dt$ i.e. when the integral is taken over a δ-neighbourhood of each point x_0, t^0 where δ is sufficiently small, but independent of a small parameter ε of the system $\bar{u}(x_0,t^0) = \lim\limits_{\delta\to 0}\lim\limits_{\varepsilon\to 0} \int_{\delta_{x_0,t^0}} u_\varepsilon(x,t)\,dx\,dt$

This is equivalent to the weak limit

$$\int \phi(x,t)u_\varepsilon(x,t)\,dx\,dt \xrightarrow[\varepsilon\to 0]{} \int \phi(x,t)\bar{u}(x,t)\,dx\,dt$$

as $\varepsilon \to 0$ for small $\phi(x,t) \in L_2(R_+, R^n)$. Obviously, such averaging satisfies all the assumptions of Reynolds averaging, in particular,

$$\overline{\partial u_\varepsilon(x,t)/\partial x} = \partial\bar{u}(x,t)/\partial x, \quad \overline{\Psi_\varepsilon(x,t)\bar{u}(x,t)} = \bar{\Psi}(x,t)\bar{u}(x,t).$$

If we take as a small parameter the inverse value of the root from the Reynolds number, then under this assumption the Navier-Stokes equation yields a chain of Reynolds equations with zero viscosity. We present here an example when such averaging and thus the Reynolds equation itself will posess the burst incorrectness. Physically this means that the averaged velocity will be as sensitive to changes of initial conditions as the instant velocity.

As an example we here consider a smoothly curved tube of constant round section (or a coaxial tube the section of which is a ring of constant radius), and the tube bending can vary in time. We need the last assumption in order to be able to carry out some experiments. An incompressible conducting fluid with large magnetic Re_M and hydrodynamic Re Reynolds numbers.

The initial "fluctuation" velocity and magnetic field oscillate along the radius with "period" $\approx 1/\sqrt{Re_M}$, the magnetic field is perpendicular to the section and sufficiently strong.

We can obtain this situation in an experiment if, on a sufficiently large interval, we have some fluid in a long straight tube in which there are small thin coaxial tubes inserted into each other, and the radiuses of these tubes differ by $\approx 1/\sqrt{\mathrm{Re}_M}$. Then if this part of the tube is sufficiently long, the oscillations of velocity and magnetic field which we want to obtain appear in the free part of the tube and we begin to bend this tube slowly and smoothly. This system posseses the burst incorrectness with respect to a small and smoothly varying angle of rotation of the tube. A small analytic adding to the initial condition appears when we begin to bend the tube, and this adding yields essentially large change of the averaged velocity in a small time. And the obtained vortexes will be very narrow across the tube and very long along the tube.

2. Consider the following initial-boundary problem for equations of magneto-hydrodynamics for incompressible fluid [1.2]

$$\frac{\partial u}{\partial t} + < u, \nabla > u = -\nabla\left(P + \frac{|H|^2}{8\pi}\right) + \frac{1}{4\pi} < H, \nabla > H + \varepsilon^2 \Delta u$$

(1)
$$\frac{\partial H}{\partial t} = \mathrm{rot}\,(u \times H) + \varepsilon^2 \nu \Delta H,$$

$$\mathrm{div}\, u = 0, \quad \mathrm{div}\, H = 0$$

(2)
$$u|_{t=0} = V^0\big(\Phi^0(x)/\varepsilon, x\big) + \varepsilon u_1^0(x),$$

$$H|_{t=0} = \mathcal{H}^0\left(\Phi^0(x)/\varepsilon, x\right) + \varepsilon H_1^0(x),$$

(3)
$$u|_{\partial\Omega} = 0, \quad \varepsilon\mathrm{rot}\, H|_{\partial\Omega} = J(x, t, \varepsilon)$$

Here $x \in \Omega \subset R^3$, the boundary $\partial\Omega$ is assumed to be smooth, $u = (u_1, u_2, u_3)$ is the velocity, p is the pressure, $H = (H_1, H_2, H_3)$ is the intensity of the magnetic field, $1/\nu$ is the conductivity coefficient normalized by the Reynolds number Re, Re $\sim \mathrm{Re}_M \gg 1$, $\varepsilon = 1/\sqrt{\mathrm{Re}}$ is a small parameter, $\Phi^0(x), u_1^0(x), H_1^0(x)$ and $V^0(\tau, x), \mathcal{H}^0(\tau, x), J(x, t, \varepsilon), \varepsilon \in [0, 1)$ are given smooth functions, $V^0(\tau, x), \mathcal{H}^0(\tau, x)$ are 2π-periodic in τ as $\tau \to \infty$, the angle brackets denote the scalar product of vectors.

We assume that the divergence of $u|_{t=0}$ and $H|_{t=0}$ is equal to zero and

$$\Phi^0|_{\partial\Omega} = 0, < \nabla\Phi^0, \bar{\mathcal{H}}^0 >= 0, < \nabla\Phi^0, J(x, t, 0) >= 0,$$

The solution (asymptotic as $\varepsilon \to 0$) describing the coherent MHD - structure was constructed in [3] (similar asymptotics for equations of gas and fluid dynamics see in [4-7]). This asymptotic solution has the form

(4)
$$u = V\left(\frac{\Phi(x, t, \varepsilon)}{\varepsilon}, x, t, \varepsilon\right), \quad H = \mathcal{H}\left(\frac{\Phi(x, t, \varepsilon)}{\varepsilon}, x, t, \varepsilon\right),$$

$$P = \mathcal{P}\left(\frac{\Phi(x, t, \varepsilon)}{\varepsilon}, x, t, \varepsilon\right)$$

where $\Phi(x,t,\varepsilon)$ is a scalar function smooth for $\varepsilon \in [0,1)$, and $V(\tau,x,t,\varepsilon)$, $\mathcal{H}(\tau,x,t,\varepsilon)$, $\mathcal{P}(\tau,x,t,\varepsilon)$ are functions smooth in $\varepsilon \in [0,1)$ and 2π-periodic in τ as $\tau \to \infty$. The coefficients of the expansion (4) into the Taylor series with respect to ε are denoted by the lower index, for example,

$$\Phi(x,t,\varepsilon) = \Phi_0(x,t) + \varepsilon\Phi_1(x,t) + \ldots,$$
$$V(\tau,x,t,\varepsilon) = V_0(\tau,x,t) + \varepsilon V_1(\tau,x,t) + \ldots.$$

It is proved that the leading term $V_0(\tau,x,t)$, $\mathcal{H}_0(\tau,x,t)$, $\mathcal{P}_0(\tau,x,t)$, $\tau = \Phi_0/\varepsilon + \Phi_1$ of the expansion (4) satisfies the problem

$$\frac{\partial V_0}{\partial t} + <V_0, \nabla> V_0 = \frac{\partial V_0}{\partial \tau} \int_0^\tau \operatorname{div} a_0 d\tau' - \frac{1}{4\pi}\frac{\partial \mathcal{H}_0}{\partial \tau}\left(\int_0^\tau \operatorname{div} h_0 d\tau' - c\right)$$

$$- \nabla\Phi_0 \Xi - \nabla P_m + \frac{1}{4\pi} <\mathcal{H}_0, \nabla> \mathcal{H}_0 + |\nabla\Phi_0|^2\frac{\partial^2 V_0}{\partial \tau^2},$$

$$(5) \quad \frac{\partial \mathcal{H}_0}{\partial t} = \frac{\partial}{\partial \tau}\left\{\mathcal{H}_0\int_0^\tau \operatorname{div} a_0 d\tau' - V_0\left(\int_0^\tau \operatorname{div} h_0 d\tau' - c\right)\right\}$$

$$+ \operatorname{rot}(V_0 \times \mathcal{H}_0) + |\nabla\Phi_0|^2\nu\frac{\partial^2 \mathcal{H}_0}{\partial \tau^2}$$

$$\frac{\partial \Phi_0}{\partial t} + <u_0, \nabla\Phi_0> = 0, \quad \frac{\partial c}{\partial t} + \operatorname{div}\left(cV_0 + \nu|\nabla\Phi|^2\frac{\partial \mathcal{H}_0}{\partial \tau}\Big|_{\tau=0}\right) = 0,$$

$$\operatorname{div} u_0 = 0, \quad \operatorname{div} \mathcal{H}_0 = 0, \quad \frac{\partial P_m}{\partial \tau} = 0,$$

$$(6) \quad V_0\Big|_{t=0} = V^0(\tau,x), \quad \mathcal{H}_0\Big|_{t=0} = \mathcal{H}^0(\tau,x),$$

$$\Phi\Big|_{t=0} = \Phi^0(x), \quad c\Big|_{t=0} = <\nabla\Phi^0, H_1^0(x)>,$$

$$(7) \quad V_0\Big|_{\tau=0,x\in\partial\Omega} = 0, \quad \nabla\Phi_0 \times \frac{\partial \mathcal{H}_0}{\partial \tau}\Big|_{\tau=0,x\in\partial\Omega} = J(x,t,0)$$

where

$$P_m = P_0 + |\mathcal{H}_0|^2/8\pi, \quad u_0 = \bar{V}_0, \quad a_0 = V_0 - u_0$$
$$H_0 = \bar{\mathcal{H}}_0, \quad h_0 = \mathcal{H}_0 - H_0$$
$$\Xi = \frac{1}{|\nabla\Phi_0|^2}\left((a_{0j} + 2u_{0j})a_{0j} - \frac{1}{4\pi}\mathcal{H}_{0j}\mathcal{H}_{0j} - \overline{a_{0j}a_{0j}} + \frac{1}{4\pi}\overline{\mathcal{H}_{0j}h_{0j}}\right)\frac{\partial^2\Phi}{\partial x_i\partial x_j},$$

and the summing is taken over repeating indeces.

The vectors $a_0(\tau, x, t)$ and $\mathcal{H}_0(\tau, x, t)$ satisfy the conditions

(8)
$$< \nabla \Phi_0, a_0 >= 0, \quad < \nabla \Phi_0, \mathcal{H}_0 >= 0.$$

The problem (5)-(7) defines uniquely $\Phi_0(x, t)$ and the functions $V_0(\tau, x, t)$, $\mathcal{H}_0(\tau, x, t)$ which are 2π-periodic as $\tau \to \infty$. However a correction to the phase Φ_0 depends on the terms of higher powers in the asymptotic expansion (4) and, in particular, on a smooth perturbation of the initial velocity $u_1^0(x)$. This fact means that the solution of problem (1)-(3) is asymptotically ill posed (see also [3-6]). Meanwhile, the averaged solution u_0, H_0, $P_0 = \bar{P}_0$ just as for equations of gas dynamics [4,5] independent of small perturbations of the velocity. To prove this statement it is sufficient to consider the Reynolds equation

(9)
$$\frac{\partial u_0}{\partial t} + < u_0, \nabla > u_0 + \nabla P_0 + \frac{1}{4\pi} H_0 \times \mathrm{rot}\, H_0 = G_1$$
$$\frac{\partial H_0}{\partial t} = \mathrm{rot}\,(u_0 \times H_0) + \mathrm{rot}\, G_2, \mathrm{div}\, u_0 = 0, \mathrm{div}\, H_0 = 0,$$
$$u_0|_{t=0} = \bar{V}^0, H_0|_{t=0} = \bar{\mathcal{H}}^0, u_0^{\perp}|_{\partial \Omega} = 0$$

and to prove that the Reynolds tensions G_1, G_2 are independent of shifts of the argument τ of the functions $V_0, \mathcal{H}_0$. Here $u_0^{\perp}$ is the normal to $\partial \Omega$ component of u_0

$$G_{1i} = -\left\{ \frac{\partial}{\partial x_j}\left(\overline{a_{0_i} a_{0_j}} - \frac{1}{4\pi}\overline{h_{0_i} h_{0_j}}\right) + \frac{1}{8\pi}\frac{\partial}{\partial x_i}\overline{(h_{0_j})^2}\right\}, \quad G_2 = \overline{(a_0 \times h_0)}.$$

We also note that the asymptotic solution (4) describes the so called effect of turbulent dynamo [1], i.e. the effect of generating the averaged magnetic field of order $O(1)$ by hydrodynamic oscillations in a finite interval of time. Actually, for $\mathcal{H}^0 = 0$ and sufficiently small t, we have

$$\frac{\partial c}{\partial t} + \mathrm{div}\,(cV^0|_{\tau=0}) = O(t), \quad \frac{\partial h_0}{\partial t} = c\frac{\partial a^0}{\partial \tau} + O(t)$$
$$c\Big|_{t=0} =< \nabla \Phi^0, H_1^0(x) >, \quad h_0\Big|_{t=0} = 0.$$

Thus, $c \neq 0$ for small smooth normal to $\partial \Omega$ perturbations of the field and the expression $c\partial V_0/\partial \tau$ in the right-hand side of (5) acts as a source of the oscillating field h_0. In its turn, the second equation in (9) shows that the interaction of the oscillating velocity a_0 and the field h_0 generate an averaged field H_0.

In the particular case $V^0 - \bar{V}^0 \to 0$, $\mathcal{H}^0 - \bar{\mathcal{H}}^0 \to 0$ as $\tau \to \infty$, i.e. when rapid perturbations of initial conditions $V^0, \mathcal{H}^0$ are concentrated in a small neighbourhood of the boundary $\partial \Omega$, the asymptotic solution (4) is an asymptotic of boundary layer type, In this case, the functions G_1, G_2 are equal to zero, the equations (7) are MHD Prandtl equations, and the effects of asymptotic illposedness mentioned above disappear.

3. The coherent structure in (4) in the general case is two-dimensional since, by conditions (8), the vectors a_0 and $\mathcal{H}_0$ lie in the plane orthogonal to $\nabla\Phi$

Consider the problem about existence of one-dimensional coherent structure. A trivial example of such structure is the solution of two-dimensional problem (1)-(3). In order to consider a three-dimensional situation we introduce in the system (7) the Lagrangian coordinate system. Suppose $x_0^1 = \Phi^0(x)$, $x_0^k = x_0^k(x)$, $k = 2, 3$ are smooth functions which are local coordinates in Ω and x_0^2, x_0^3 are the initial coordinates on $\partial\Omega$. We introduce the functions $x^j = x^j(x, t)$ as the solution of the following problem

$$\frac{\partial x^j}{\partial t} + <u_0, \nabla x^j> = 0, \quad x^j\Big|_{t=0} = x_0^j(x), \; j = 1, 2, 3$$

We can see that the functions x^j are also local coordinates in $\bar{\Omega}$ for a smooth function u_0.

We denote by $x_i = x_i(x^1, x^2, x^3, \Phi)$ the solution of the following equations $x^j = x^j(x_1, x_2, x_3, t)$; then $e^j = \nabla x^j$, $e_j = (\partial x_1/\partial x^j, \partial x_2/\partial x^j, \partial x_3/\partial x^j)$ are local basis vectors, $g^{jk} = <e^j, e^k>$, $g_{jk} = <e_j, e_k>$ are the coefficients of matrix tensors, $g = \det g_{jk}$; F^j are the coefficients of decomposition of the vector F with respect to the basis (e_1, e_2, e_3), $F = e_j F^j$. We also denote by $F_{,k}^j$ the covariant derivative

$$F_{,k}^j = \frac{\partial F^j}{\partial x^k} + \Gamma_{ki}^j F^i,$$

where $\Gamma_{ki}^j = <e^j, \partial e^k/\partial x^i>$ are the Christoffel symbols of the second order and we preserve the notations of (4) for the functions in the variables τ, x^j, t. We note that, by conditions (8), $\mathcal{H}_0^1 = 0$, $V_0^1 = u_0^1$. We pass to the Lagrangian system of coordinates in the equations (5). After simple calculations we obtain the following simple statement.

LEMMA 1. *The system of equations (5) in Lagrangian coordinates has the following form*

$$(10) \qquad \frac{\partial u_0^1}{\partial t} + u_0^i u_{0,i}^1 + g^{1i}\frac{\partial P_m}{\partial x^i} - \frac{1}{4\pi}\Gamma_{ij}^1 H_0^i H_0^j = G_1^1,$$

$$(11) \quad \frac{\partial V_0^k}{\partial t} + V_0^i u_{0,i}^k + a_0^i V_{0,i}^k + g^{ki}\frac{\partial P_m}{\partial x^i} =$$

$$\frac{\partial V_0^k}{\partial \tau}\int_0^\tau \operatorname{div} a_0 d\tau' - \frac{1}{4\pi}\frac{\partial h_0^k}{\partial \tau}\Big(\int_0^\tau \operatorname{div} h_0 d\tau' - c\Big) - g^{k1}\Xi +$$

$$\frac{1}{4\pi}\mathcal{H}_0^i \mathcal{H}_{0,i}^k + g^{11}\frac{\partial^2 V_0^k}{\partial \tau^2},$$

$$(12) \quad \frac{\partial \mathcal{H}_0^k}{\partial t} + \frac{1}{\sqrt{|g|}}\frac{\partial}{\partial x^i}\sqrt{|g|}(a_0^i \mathcal{H}_0^k - a_0^k \mathcal{H}_0^i)$$

$$= \frac{\partial}{\partial \tau}\Big\{\mathcal{H}_0^k\int_0^\tau \operatorname{div} a_0 d\tau' - a_0^k\Big(\int_0^\tau \operatorname{div} h_0 d\tau' - c\Big)\Big\} + g^{11}\nu\frac{\partial^2 \mathcal{H}_0^k}{\partial \tau^2},$$

$$(13) \qquad \frac{\partial c}{\partial t} + \mathrm{div}\left(ca_0 + \nu g^{11}\frac{\partial h_0}{\partial \tau}|_{\tau=0}\right) = 0,$$

$$(14) \qquad \mathrm{div}\, u_0 = 0, \quad \mathrm{div}\, H_0 = 0, \quad \frac{\partial P_m}{\partial \tau} = 0$$

where $k = 2, 3$

$$P_m = \mathcal{P}_0 + |\mathcal{H}_0|^2/8\pi, \quad |F|^2 = g_{jk}F^jF^k,$$

$$\mathrm{div}\, F = \frac{1}{\sqrt{|g|}}\partial/\partial x^j(\sqrt{|g|}F^j)$$

$$\Xi = \frac{1}{g''}(-2a_0^i u_{0,i}^1 + \Gamma_{ij}^1(\mathcal{H}_0^i\mathcal{H}_0^j/4\pi - a_0^i a_0^j + \overline{a_0^i a_0^j} - \overline{\mathcal{H}_0^i\mathcal{H}_0^j/4\pi}))$$

$$G_1^1 = -\Gamma_{ij}^1\overline{(a_0^i a_0^j - h_0^i h_0^j/4\pi)}.$$

Suppose the coherent structure (2) is one-dimensional and parallel to the vector e_2 for $t = 0$. The structure remains one-dimensional for $t > 0$ if, besides of (8), the conditions

$$(15) \qquad < e^3, a_0 >= 0, \quad < e^3, \mathcal{H}_0 >= 0$$

hold uniquely in t.

LEMMA 2. *In order (15) hold, it is necessary and sufficient that*

$$(16) \qquad \begin{aligned} g^{11}u_{0,2}^3 - g^{31}u_{0,2}^1 &= 0 \\ g^{11}\Gamma_{22}^3 - g^{31}\Gamma_{22}^1 &= 0 \\ (g^{11}g^{3j} - g^{31}g^{1j})\frac{\partial P_m}{\partial x^j} &= 0 \end{aligned}$$

Necessity. By (15), we have $a_0^3 = 0$, $\mathcal{H}_0^3 = 0$. This and Lemma 1 yield that, for $k = 3$, (12) holds identically and (11) separates into conditions (16) and the following equation for the averaged u_0^3

$$(17) \qquad \frac{\partial u_0^3}{\partial t} + u_0^i u_{0,i}^3 + g^{3i}\frac{\partial P_m}{\partial x^i} - \frac{1}{4\pi}\Gamma_{22}^3(H_0^2)^2 = G_1^3$$
$$G_1^3 = \Gamma_{22}^3\{(h_0^3)^2/4\pi - (a_0^2)^2\}.$$

Sufficiency. In order to prove the second statement, it is sufficient to note that, for $k = 3$, under (16), the equations (11), (12) form a homogeneous system which has only a trivial solution for zero initial conditions.

The conditions (16) are very strong and can hold only in special cases of symmetry of the domain Ω and the initial data. We consider some examples of one-dimensional structures which are rapidly oscillating analogs of equilibrium configurations.

of coordinates for $t = 0$, $r = \sqrt{x_1^2 + x_2^2}$, $r_0 = $ const. Suppose the initial functions have the form

$$V^0 = e_1^0 u^{01}(r, x_3) + e_2^0 V^{02}(\frac{\Phi^0}{\varepsilon}, r, x_3)$$

(18)

$$\mathcal{H}^0 = e_2^0 \mathcal{H}^{02}(\frac{\Phi^0}{\varepsilon}, r, x_3)$$

where $e_1^0 = -i_1 \cos\theta - i_2 \sin\theta$, $e_2^0 = -i_3$, i_j the orths of the Cartesian system of coordinates, $\mathcal{H}^{02} = b/r$, $b = $ const or $b = b(r)$ for $u^{01} = 0$.

By passing to Lagrangain coordinates, we see

$$x^1 = x^1(r, x_3, t), \quad x^2 = x^2(r, x_3, t), \quad x^3 = \theta,$$

$$e_1 = (i_1 \frac{\partial x^2}{\partial x_3} \cos\theta + i_2 \frac{\partial x^2}{\partial x_3} \sin\theta - i_3 \frac{\partial x^2}{\partial r})/\Delta|_{x_i = x_i(x^j, t)}$$

$$e_2 = (i_1 \frac{\partial x^1}{\partial x_3} \cos\theta + i_2 \frac{\partial x^2}{\partial x_3} \sin\theta - i_3 \frac{\partial x^1}{\partial r})/\Delta|_{x_i = x_i(x^j, t)}$$

$$\Delta = \frac{\partial x^1}{\partial r} \frac{\partial x^2}{\partial x_3} - \frac{\partial x^1}{\partial x_3} \frac{\partial x^2}{\partial r}.$$

After the coefficients of metric tensors and the Christoffel symbols are calculated, we obtain $g_{13} = g_{23} = g^{13} = g^{23} = 0$, $\Gamma_{13}^l = \Gamma_{23}^l = 0$, $l = 1, 2$, $\Gamma_{ij}^3 = 0$ except the cases $i = 1, j = 3$ and $i = 2, j = 3$. The conditions of Lemma 2 are satisfied and the leading term of the asymptotic solution has the form

$$V^0 = e_1 u_0^1(x^1, x^2, t) + e_2 V_0^2(\tau, x^1, x^2, t)$$

(19)

$$\mathcal{H}_0 = e_2 \mathcal{H}_0^2(\tau, x^1, x^2, t)$$

where $H_0^2 = b/\sqrt{|g|}$, the other functions can be defined by equations (10), (13) and the equations (11), (12) which, in our case, have the form

(20) $$\frac{\partial V_0^2}{\partial t} + u_0^i u_{0,i}^2 + a_0^2(2u_{0,2}^2 + a_{0,2}^2) + g^{2i}\frac{\partial P_m}{\partial x^i}$$

$$= \frac{\partial a_0^2}{\partial \tau}\int_0^\tau \text{div } a_0 d\tau' - \frac{1}{4\pi}\frac{\partial h_0^2}{\partial \tau}\left(\int_0^\tau \text{div } h_0 d\tau' - c\right) - g^{21}\Xi$$

$$+ \frac{1}{4\pi}\mathcal{H}_0^2\mathcal{H}_{0,2}^2 + g^{11}\frac{\partial^2 V_0^2}{\partial \tau^2}$$

(21) $$\frac{\partial h_0^2}{\partial t} = \frac{\partial}{\partial \tau}\{\mathcal{H}_0^2\int_0^\tau \text{div } h_0 d\tau' - a_0^2(\int_0^\tau \text{div } h_0 d\tau' - c)\} + g^{11}\nu\frac{\partial^2 h_0^2}{\partial \tau^2}$$

where $\text{div } a_0 = \frac{1}{\sqrt{|g|}}\frac{\partial}{\partial x^2}(\sqrt{|g|}a_0^2)$, $\text{div } h_0 = \frac{1}{\sqrt{|g|}}\frac{\partial}{\partial x^2}(\sqrt{|g|}h_0^2)$, $\mathcal{H}_{0,2}^2 = \frac{\partial \mathcal{H}_0^2}{\partial x^2} + \Gamma_{22}^2\mathcal{H}_0^2$.

Example 2. Z-pinch. Suppose the initial data are independent of the height $z = x_3$. We denote by $x_2^1 = \Phi^0 = r_0 - r$, $x_0^2 = \theta$, $x_0^3 = x_3$ a cylindrical system of coordinates. Consider the initial conditions of the form (18) where

$$e_1^0 = i_1 \frac{1}{\Delta}\left(\frac{\partial x^2}{\partial \theta}\cos\theta + r\frac{\partial x^2}{\partial r}\sin\theta\right) + i_2\frac{1}{\Delta}\left(\frac{\partial x^2}{\partial \theta}\sin\theta - r\frac{\partial x^2}{\partial r}\cos\theta\right)|_{t=0}$$

$$e_2^0 = i_1 \frac{1}{\Delta}\left(\frac{\partial x^1}{\partial \theta}\cos\theta + r\frac{\partial x^1}{\partial r}\sin\theta\right) + i_2\frac{1}{\Delta}\left(\frac{\partial x^1}{\partial \theta}\sin\theta - r\frac{\partial x^1}{\partial r}\cos\theta\right)|_{t=0}$$

$$\Delta = \frac{\partial x^1}{\partial r}\frac{\partial x^2}{\partial \theta} - \frac{\partial x^1}{\partial \theta}\frac{\partial x^2}{\partial r}$$

$$H^{02} = b\Delta|_{t=0}/r, \quad b = \text{const, or } b = b(r) \text{ for } u^{01} = 0.$$

By passing to Lagrangian coordinates, we obtain $x^1 = x^1(r,\theta,t)$, $x^2 = x^2(r,\theta,t)$, $x^3 = x_3$, $e_3 = i_3$. And $g_{i3} = g^{i3} = 0$, $i = 1,2$, $\Gamma_{j3}^l = 0$ for $l = 1,2$, $j = 1,2,3$, $\Gamma_{ij}^3 = 0$, $i,j = 1,2,3$. The conditions of Lemma 2 hold and we have formulas (19)-(21) for the leading term of the asymptotic solution.

Example 3. Torus configuration. . Denote

$$\phi = \arctan(x_2/x_1),$$

$$r = \frac{1}{2}\left((\sqrt{(x_1)^2 + (x_2)^2} + a)^2 + (x_3)^2\right)$$

$$-\frac{1}{2}\ln((\sqrt{(x_1)^2 + (x_2)^2} - a)^2 + (x_3)^2),$$

$$\sigma = \frac{i}{2}\ln((x_1)^2 + (x_2)^2) + (x_3 - ia)^2) - \frac{i}{2}\ln((x_1)^2 + (x_2)^2) + (x_3 + ia)^2),$$

$-\pi \le \sigma \le \pi$, $0 \le \phi < 2\pi$, $a > 0$ is a certain constant, and let $x_0^1 = \Phi^0(r,\sigma)$, $x_0^2 = x_0^2(r,\sigma)$, $x_0^3 = \phi$ be local coordinates for $t = 0$. Consider the initial conditions of the form (18) where

$$e_1^0 = i_1\Delta^{-1}\left(\frac{\partial f_1}{\partial r}\frac{\partial x^2}{\partial \sigma} - \frac{\partial f_1}{\partial \sigma}\frac{\partial x^2}{\partial r}\right)\cos\phi + i_2\Delta^{-1}\left(\frac{\partial f_1}{\partial r}\frac{\partial x^2}{\partial \sigma} - \frac{\partial f_1}{\partial \sigma}\frac{\partial x^2}{\partial r}\right)\sin\phi +$$
$$i_3\Delta^{-1}\left(\frac{\partial f_2}{\partial r}\frac{\partial x^2}{\partial \sigma} - \frac{\partial f_2}{\partial \sigma}\frac{\partial x^2}{\partial r}\right)|_{t=0},$$

$$e_2^0 = i_1\Delta^{-1}\left(\frac{\partial f_1}{\partial r}\frac{\partial x^1}{\partial \sigma} - \frac{\partial f_1}{\partial \sigma}\frac{\partial x^1}{\partial r}\right)\cos\phi + i_2\Delta^{-1}\left(\frac{\partial f_1}{\partial r}\frac{\partial x^1}{\partial \sigma} - \frac{\partial f_1}{\partial \sigma}\frac{\partial x^1}{\partial r}\right)\sin\phi +$$
$$i_3\Delta^{-1}\left(\frac{\partial f_2}{\partial r}\frac{\partial x^1}{\partial \sigma} - \frac{\partial f_2}{\partial \sigma}\frac{\partial x^1}{\partial r}\right)|_{t=0},$$

$$f_1 = \frac{a\sinh r}{\cosh r - \cos\sigma}, \quad f_2 = \frac{a\sin\sigma}{\cosh r - \cos\sigma}, \quad \Delta = \frac{\partial x^1}{\partial \sigma}\frac{\partial x^2}{\partial r} - \frac{\partial x^1}{\partial r}\frac{\partial x^2}{\partial \sigma}.$$

By passing to Lagrangian coordinates, we obtain $x^1 = x^1(r,\sigma,t)$, $x^2 = x^2(r,\sigma,t)$, $x^3 = \phi$. And $g_{i3} = g^{i3} = 0$, $i = 1,2$, $\Gamma_{i3}^l = 0$, $l = 1,2$, $i = 1,2$, $\Gamma_{jj}^3 = 0$, $j = 1,2,3$, $\Gamma_{12}^3 = 0$.

The conditions of Lemma 2 hold and the leading term of the asymptotic solution satisfies the formulas (19)-(21).

4. We now begin to consider a problem about stability of the asymptotic solution constructed with respect to rapidly oscillating preturbations.

Suppose

(22)
$$u|_{t=0} = V^0\left(\frac{\Phi^0(x)}{\varepsilon}, x\right) + \delta U^0\left(\frac{\Phi^0(x)}{\varepsilon}, \frac{\Psi^0(x)}{\varepsilon}, x\right),$$
$$H|_{t=0} = \mathcal{H}^0\left(\frac{\Phi^0(x)}{\varepsilon}, x\right) + \delta B^0\left(\frac{\Phi^0(x)}{\varepsilon}, \frac{\Psi^0(x)}{\varepsilon}, x\right)$$

where V^0, $\mathcal{H}^0$, Φ^0 are functions from (2), $U^0(\tau, \eta, x)$, $B^0(\tau, \eta, x)$, are periodic in τ as $\tau \to \infty$ and 2π-periodic in η, $\Psi^0(x) \in C^\infty$.

A self-similar asymptotic solution of problem (1),(22) will be written in the form similar to (22)

(23)
$$u = V_0(\tau, x, t) + \delta U(\tau, \eta, x, t) + \dots$$
$$H = \mathcal{H}_0(\tau, x, t) + \delta B(\tau, \eta, x, t) + \dots$$

where $\tau = \Phi_0(x, t)/\varepsilon + \Phi_1(x, t)$, V_0, $\mathcal{H}_0$ are the leading terms of one-phase asymptotics, $\eta = \Psi(x, t)/\varepsilon$, $\Psi \in C^\infty$, U, B are smooth functions 2π-periodic in η and τ as $\tau \to \infty$.

By substituting (23) into (1), we see that the leading term of asymptotic expansion satisfies the following equations

(24) $\left(\dfrac{\partial \Psi}{\partial t} + \langle V_o, \nabla \Psi \rangle\right)\dfrac{\partial U}{\partial \eta} + \langle U, \nabla \Phi_0\rangle\dfrac{\partial V_0}{\partial \tau} + \nabla \Phi_0\dfrac{\partial P_m}{\partial \tau} + \nabla \Psi\dfrac{\partial P_m}{\partial \eta} =$

$$\frac{1}{4\pi}\langle B, \nabla \Phi_0\rangle\frac{\partial \mathcal{H}_0}{\partial \tau} + \frac{1}{4\pi}\langle \mathcal{H}_0, \nabla \Psi\rangle\frac{\partial B}{\partial \eta}$$

$$\left(\frac{\partial \Psi}{\partial t} + \langle V_o, \nabla \Psi\rangle\right)\frac{\partial B}{\partial \eta} + \langle U, \nabla \Phi_0\rangle\frac{\partial \mathcal{H}_0}{\partial \tau} = \langle B, \nabla \Phi_0\rangle\frac{\partial \mathcal{H}_0}{\partial \tau} + \langle \mathcal{H}_0, \nabla \Psi\rangle\frac{\partial B}{\partial \eta}$$

$$\frac{\partial}{\partial \tau}\langle U, \nabla \Phi_0\rangle + \frac{\partial}{\partial \eta}\langle U, \nabla \Psi\rangle = 0, \frac{\partial}{\partial \tau}\langle B, \nabla \Phi_0\rangle + \frac{\partial}{\partial \eta}\langle B, \nabla \Psi\rangle = 0$$

We introduce a new variable $\eta_1 = \eta - q\tau$ where $q = \langle \nabla \Phi_0, \nabla \Psi\rangle/|\nabla \Phi_0|^2$ and denote

$$\chi(\tau, \eta_1, x, t) = \langle U(\tau, \eta_1 + q\tau, x, t), \nabla \Phi_0\rangle,$$
$$\varkappa(\tau, \eta_1, x, t) = \langle U(\tau, \eta_1 + q\tau, x, t), k\rangle$$
$$\alpha(\tau, \eta_1, x, t) = \langle B(\tau, \eta_1 + q\tau, x, t), \nabla \Phi_0\rangle,$$
$$\beta(\tau, \eta_1, x, t) = \langle B(\tau, \eta_1 + q\tau, x, t), k\rangle$$

where $k = \nabla \Psi - q\nabla \Phi_0$.

Let us simplify the system (24). After multiplying equations (24) by the vectors $\nabla\Phi_0$ and k, we obtain the following scalar equations

(25)
$$(\dot{\Psi} + a_\Psi)\frac{\partial\chi}{\partial\eta_1} + |\nabla\Phi_0|^2\frac{\partial P_m}{\partial\tau} = \frac{1}{4\pi}\mathcal{H}_\Psi\frac{\partial\alpha}{\partial\eta_1}$$

(26)
$$(\dot{\Psi} + a_\Psi)\frac{\partial\varkappa}{\partial\eta_1} + \chi\frac{\partial a_\Psi}{\partial\tau} + |k|^2\frac{\partial P_m}{\partial\eta_1} = \frac{\alpha}{4\pi}\frac{\partial\mathcal{H}_\Psi}{\partial\tau} + \frac{\mathcal{H}_\Psi}{4\pi}\frac{\partial\beta}{\partial\eta_1}$$

(27)
$$(\dot{\Psi} + a_\Psi)\frac{\partial\alpha}{\partial\eta_1} = \mathcal{H}_\Psi\frac{\partial\chi}{\partial\eta_1}$$

(28)
$$(\dot{\Psi} + a_\Psi)\frac{\partial\beta}{\partial\eta_1} + \chi\frac{\partial\mathcal{H}_\Psi}{\partial\tau} = \alpha\frac{\partial a_\Psi}{\partial\tau} + \mathcal{H}_\Psi\frac{\partial\varkappa}{\partial\eta_1}$$

(29)
$$\frac{\partial\chi}{\partial\tau} + \frac{\partial\varkappa}{\partial\eta_1} = 0, \quad \frac{\partial\alpha}{\partial\tau} + \frac{\partial\beta}{\partial\eta_1} = 0$$

where $\dot{\Psi} = \frac{\partial\Psi}{\partial t} + \langle u_0, \nabla\Psi\rangle$, $a_\Psi = \langle a_0, \nabla\Psi\rangle$, $\mathcal{H}_\Psi\langle\mathcal{H}_0, \nabla\Psi\rangle$.

We note that, by means of relations (29), we can integrate the equations (27), (28) and obtain

(30)
$$(\dot{\Psi} + a_\Psi)\alpha = \mathcal{H}_\Psi\chi$$

Moreover, multiplying (24) by the vector n orthogonal to $\nabla\Phi_0$ and k, we obtain the following equations for the projections $U_n = \langle U, n\rangle$ and $B_n = \langle B, n\rangle$

$$\frac{\partial}{\partial\eta_1}\left((\dot{\Psi} + a_\Psi)U_n - \frac{1}{4\pi}\mathcal{H}_\Psi B_n\right) = \frac{\alpha}{4\pi}\frac{\partial\langle\mathcal{H}_0, n\rangle}{\partial\tau} - \chi\frac{\partial\langle a_0, n\rangle}{\partial\tau}$$

$$\frac{\partial}{\partial\eta_1}\left((\dot{\Psi} + a_\Psi)B_n - \mathcal{H}_\Psi U_n\right) = \alpha\frac{\partial\langle a_0, n\rangle}{\partial\tau} - \chi\frac{\partial\langle\mathcal{H}_0, n\rangle}{\partial\tau}$$

Further, we shall consider certain one-dimensional structures in two situations $\langle\nabla\Psi, a_0\rangle = 0$ and $\nabla\Psi\|a_0$.

Suppose the vectors a_0 and $\nabla\Psi$ are orthogonal. Then $a_\Psi = \mathcal{H}_\Psi = 0$, and (25),(26),(30) yield the equation for the phase ϕ

$$\dot{\Psi} = 0$$

as well as the condition that the magnetic pressure $P_m = P_m(x, t)$ is independent of τ and η_1. Thus the asymptotic solution is stable in linear approximation with respect to these perturbations.

Suppose the vectors a_0 and $\nabla\Psi$ are parallel. We note that $\eta_1 = \eta$ in this situation. We exclude the functions $\varkappa, \alpha, \beta, P_m$ from the system (25), (26), (29), (30) and represent the function χ in the form $\chi = (\dot{\Psi} + a_\Psi)\phi(\tau, \eta, x, t)$. Then we obtain the following problem about eigenvalues which allows to define Ψ and ϕ

(31)
$$|\nabla\Phi_0|^2\frac{\partial}{\partial\tau}\left(\lambda\frac{\partial\phi}{\partial\tau}\right) + |\nabla\Psi|^2\lambda\frac{\partial^2\phi}{\partial\eta^2} = 0$$

$$\frac{\partial^i\phi}{\partial\eta^i}\Big|_{\eta=0} = \frac{\partial^i\phi}{\partial\eta^i}\Big|_{\eta=2\pi}, \quad \lim_{s\to\infty}\left(\frac{\partial^i\phi}{\partial\tau^i}\cdot\Big|_{\tau=\zeta} - \frac{\partial^i\phi}{\partial\tau^i}\Big|_{\tau=\zeta+2\pi}\right) = 0$$

where $i = 0, 1$, $\lambda = (\dot{\Psi} + a_\Psi)^2 - (\mathcal{H}_\Psi)^2/4\pi$. On the other hand, (31) easily yields the following relations

$$\dot{\Psi} = -\frac{f_1}{f_0} \pm \sqrt{\left(\frac{f_1}{f_0}\right)^2 + \frac{f_2}{f_0}},$$

$$f_0 = \frac{1}{4\pi^2} \lim_{\zeta \to \infty} \int_0^{2\pi} \int_\zeta^{\zeta+2\pi} I \, d\tau \, d\eta,$$

(32)
$$f_1 = \frac{1}{4\pi^2} \lim_{\zeta \to \infty} \int_0^{2\pi} \int_\zeta^{\zeta+2\pi} I a_\Psi \, d\tau \, d\eta,$$

$$f_2 = \frac{1}{2\pi} \lim_{\zeta \to \infty} \int_\zeta^{\zeta+2\pi} (\mathcal{H}_\Psi)^2/4\pi - (a_\Psi)^2 \, d\tau,$$

$$I = |\nabla \Phi_0|^2 \left|\frac{\partial \phi}{\partial \tau}\right|^2 + |\nabla \Psi|^2 \left|\frac{\partial \phi}{\partial \eta}\right|^2.$$

Since the coefficients f_i are real, (32) yields that an one-dimensional coherent structure is stable in linear approximation under the following assumption

$$(H_\Psi)^2 + \overline{(h_\Psi)^2} + 2 \left(\int_0^{2\pi} \overline{I a_\Psi} \, d\eta\right)^2 \Big/ \int_0^{2\pi} \overline{I} \, d\eta \geq 4\pi \overline{(a_\Psi)^2}$$

i.e. for sufficiently strong magnetic field.

We note that if $|\nabla \Psi| << 1$, i.e. if the lengths of waves of perturbating oscillation is much greater than those of basic oscillations, the relation (32) can be substituted by

(33)
$$(\dot{\Psi})^2 = \frac{1}{4\pi} \left((H_\Psi)^2 + \overline{(h_\Psi)^2}\right) - \overline{(a_\Psi)^2}$$

The Hamilton-Jacobi equation (33) corresponds to the following dispersion relation

(34)
$$\omega^2 = \frac{1}{4\pi} \left(\langle H, k \rangle^2 + \overline{\langle h, k \rangle^2}\right) - \overline{\langle a, k \rangle^2},$$

where $\omega = \dot{\Psi}$, $k = \nabla \Psi$. If the right-hand side of (34) is negative for certain k, then the perturbation is not stable and its increment of growth is equal to the square root of modulus of the right-hand side of (34). By using the formula (34), one can easily prove the following statement

LEMMA 3. *Suppose* $1 >> |k| >> \varepsilon$. *Then in order the flow is stable with respect to perturbations of form (22), it is necessary and sufficient that the vector* $\mathcal{H}_0(\tau, x, t)$ *be parallel to the vector* $a_0(\tau, x, t)$, *and the right-hand side of (34) be nonnegative for any direction of the vector* k.

5. We show that any small smooth perturbations of the boundary result in burst incorrectness

where $r_0 = \text{const}$, α is a small parameter, $\alpha \gg \varepsilon$, $f \in C_0^\infty$, $\frac{\partial f}{\partial x_3} \neq 0$.

There exist stable one-dimensional flows in a cylindrical domain with constant boundary ($\alpha = 0$). For example, we have θ-pinch of the following form

$$v = e_2 \phi \left(\frac{\Phi^0(r)}{\varepsilon}, t \right), \quad H = e_2 b$$

where $e_2 = i_3$ is a vector coinciding with the axis of cylinder, $\Phi^0 = r_0 - r$, $\phi(\tau, t)$ (a 2π-periodic scalar function in τ)is the solution of the heat equation

$$\frac{\partial \phi}{\partial t} = \frac{\partial^2 \phi}{\partial \tau^2},$$

$$b = \text{const}, \quad b^2 > 2 \int_0^{2\pi} \phi^2 \, d\tau$$

For $\alpha \neq 0$, we consider the perturbation of this flow of the following form

(36)
$$v|_{t=0} = e_2 \phi \left(\frac{\Phi^0}{\varepsilon}, 0 \right) + \delta U^0 \left(\frac{\Phi^0}{\varepsilon}, \frac{\Psi^0}{\varepsilon}, x \right)$$

$$H|_{t=0} = e_2 b + \delta B^0 \left(\frac{\Phi^0}{\varepsilon}, \frac{\Psi^0}{\varepsilon}, x \right)$$

where δ is a small parameter, $\Psi^0 \in C^\infty$, $\langle \nabla \Psi^0, \nabla \Phi^0 \rangle = 0$, $U^0(\tau, \eta, x)$; $B^0(\tau, \eta, x)$ are smooth functions 2π-periodic in τ and η. $< B^0, \nabla \Phi^0 >= 0$.

By choosing $\langle H|_{t=0}, \nabla \Phi_0 \rangle = 0$ and thus $c|_{t=0} = 0$, we exclude the possibility of arising the effect of turbulent dynamo. In order to exclude the effects of boundary layer, we restrict ourselves by the boundary condition of nonpermeability

(37)
$$v^\perp|_{r=R} = 0, \quad (\text{rot } H)^\perp|_{r=R} = 0$$

We now formulate the main result of this paper.

THEOREM. *The solution of problem* (1), (35)-(37) *is unstable, and the amplitude of perturbation increases in time* $t = O\left(\left(\frac{\varepsilon}{\alpha} \ln \frac{1}{\delta} \right)^{1/3} \right)$ *till the value* $O(1)$.

Proof. We assume that the solution of problem (1), (36), (37) with boundary which moves according to equation (35) remains stable. Then, by Lemma 3, we see that the vector $\mathcal{H}_0$ is parallel to the vector a_0. It is easy to see that then the equation (13) and the oscillating part of equation (12) are homogeneous in h_0 and ε. Taking into account that $h_0|_{t=0} = 0$, $c_0|_{t=0} = 0$, we get $h_0 \equiv 0$, $c \equiv 0$, i.e. the magnetic field does not oscillate.

Consider the behaviour of the solution for small values of time. Denote the stable solution for an unmovable boundary by the index (0), and the perturbation arising for small t, by functions without any index. Then the system of equations for variations of stable flow will have the form

(38)
$$\frac{\partial \Phi}{\partial t} + \langle u, \nabla \Phi^0 \rangle = 0$$

$$\langle a_0, \nabla \Phi \rangle + \langle a, \nabla \Phi^0 \rangle = 0, \quad \langle H_0, \nabla \Phi \rangle + \langle H, \nabla \Phi^0 \rangle = 0$$

$$\text{div } u = 0, \quad \text{div } H = 0$$

(39) $\quad \dfrac{\partial u}{\partial t} + \nabla(P + \dfrac{1}{4\pi}\langle H_0, H\rangle) = \dfrac{1}{4\pi}\langle H_0, \nabla\rangle H - \overline{\langle a_0, \nabla\rangle a} - \overline{a_0 \operatorname{div} a}$

(40) $\quad \dfrac{\partial H}{\partial t} = \langle H_0, \nabla\rangle u$

(41) $\quad \dfrac{\partial a}{\partial t} + \langle a_0, \nabla\rangle(u + a)$

$$= \dfrac{\partial a_0}{\partial \tau} \int_0^\tau \operatorname{div} a \, d\tau' - \nabla\Phi^0 \Xi + \dfrac{\partial^2 a}{\partial \tau^2} + \overline{\langle a_0, \nabla\rangle a} + \overline{a_0 \operatorname{div} a}$$

where $\Xi = -2\langle a_0, \nabla\rangle\langle u, \nabla\Phi^0\rangle - \langle \nabla\Phi^0, \langle a_0, \nabla\rangle a\rangle$. Respectively, the boundary and initial conditions have the form

(42) $\quad \begin{array}{c} \langle \operatorname{rot} h, \nabla\Phi^0\rangle|_{r=R} = 0, \quad \langle u, \nabla\Phi^0\rangle|_{r=R} = 0 \\ a|_{t=0} = 0, \quad H|_{t=0} = 0, \quad u|_{t=0} = 0, \quad \Phi|_{t=0} = 0 \end{array}$

The equations (39) and the boundary conditions (35), (42) yields that $u = O(\alpha t)$.

For small t, the functions a, Φ, H, P, u are proportional to a certain power of t. Thus the derivatives of these functions in t are more greater than the derivatives in spatial variables. By taking this into account in (40), (41), we find that the vectors a and H satisfy the equations

(43) $\quad \begin{array}{c} \dfrac{\partial a}{\partial t} + \langle a_0, \nabla\rangle u - 2\nabla\Phi^0 \langle a_0, \nabla\rangle\langle u, \nabla\Phi^0\rangle = 0 \\[2mm] \dfrac{\partial H}{\partial t} - \langle H_0, \nabla\rangle u = 0 \end{array}$

in the main approximation.

The relations (41) contradict to the assumption that the vectors a and H are parallel. In order to see this, it is sufficient to consider the equations for the second and the third components of vectors a and H in the cylindrical system of coordinates $x_2 = x_3, x^3 = \theta$)

$$\dfrac{\partial a^k}{\partial t} = -\phi\dfrac{\partial u^k}{\partial x^2}, \quad \dfrac{\partial H^k}{\partial t} = b\dfrac{\partial u^k}{\partial x^2}, \quad k = 2, 3$$

By (43), we can also see that the component of the vector a orthogonal to the vector H is equal to $O(\alpha t^2)$.

We now assume that the vector k in (34) is orthogonal to the magnetic field. Then the imaginary part of ω is equal to $O(\alpha t^2)$ and hence the increment of growth of the unstable harmonic is equal to $O(\alpha t^3)$. Thus the amplitude of perturbation becomes equal to $O(1)$ at time $t = O((\frac{\varepsilon}{\alpha} \ln \frac{1}{\delta})^{1/3})$.

In the same way we can see that the statement about burst incorrectness in a domain with moving boundary holds also for flows considered in Examples 2 and 3.

REFERENCES

[1] L.D.LANDAU AND E.M. LIFSHITZ, *Electrodynamics of continuous media*, Nauka, Moscow, 1982.

[2] B.B.KADOMTZEV, *Collective phenomena in plasm*, Nauka, Moscow, 1988.

[3] V.P.MASLOV AND G.A.OMELYANOV, *One-phase asymptotics for equations of magneto hydrodynamics under large Reynolds numbers*, Sibirsk. Mat. Z., 29 (1988), pp. 172-180. Russian

[4] V.P.MASLOV, *Coherent structures, resonances and asymptotic nonuniqueness for Navier - Stokes equations under large Reynolds numbers*, Uspekhi Mat. Nauk, 41 (1986), pp. 19-35. Russian

[5] V.P.MASLOV, *Violence of determination principle for unstationary equations of two- or three-dimensional equations of gas dynamics under sufficiently large Reynolds numbers*, Theor. Mat. Phys., 69 (1986), pp. 361-378. Russian

[6] V.P.MASLOV, *Asymptotic methods of solving pseudo differential equations*, Nauka, Moscow, 1987. Russian

[7] V.P.MASLOV, *On the situation of attractor instability in hydrodynamic turbulence*, Soviet Math. Dokl., 41 (1990), pp. 128-131.

LOSS OF STABILITY OF THE GLOBALLY UNIQUE STEADY-STATE EQUILIBRIUM AND THE BIFURCATION OF CLOSED ORBITS IN A CLASS OF NAVIER-STOKES TYPE DYNAMICAL SYSTEMS

GHEORGHE MINEA††

Abstract. We characterize the injective operators of Navier-Stokes type $Au + B(u)$ in $\mathbf{R}^3$ and also the injective operators whose linearization at each point has the spectrum in the right open half plane, such that for each right hand side f the unique stationary solution of the equation $\dot{u} + Au + B(u) = f$ be stable.

It follows that in the class of injective operators the contrary case is generic. For an injective operator in generic position we study the evolution of the attraction basin of the stationary solution, the Hopf bifurcation and the stability of the bifurcating closed orbits.

Key words. Navier-Stokes equation, Hopf bifurcation.

0. Introduction. Fifteen years ago Ciprian Foias suggested to me the study of a class of dynamical systems in the Euclidean space defined by the few axioms he was considering the essential algebraic and geometric properties of the generator of the Navier-Stokes dynamical system in a suitable real infinite dimensional Hilbert space. For quite a long time I was considering this framework too general for yielding the finer features of the solutions of these equations. Now I am thinking it is only general enough to allow the classifying entities of the generator to relieve under the action of the group of Euclidean displacements in the phase space. And my main purpose here is to point out how natural and efficient this axiomatics can be when a canonical form of the nonlinear term is available, as in the case of a 3-dimensional phase space.

We are concerned with the simplest bifurcation problem for such a dynamical system. Taking the "external forces" as parameter of bifurcation we characterize in §1 the operators for which the stationary point is unique for all values of the parameter. There, as well as in §2 and §3, we stress the fact that the canonical basis for the nonlinear term is the canonical one for the bifurcation problem. The result of §4 is intended to suggest that for great enough values of the parameter the Hopf bifurcation is supercritical and stable, while under certain circumstances the subcritical bifurcation occurs too. This stealthy kind of theft of stability is pointed out by Hopf in his basic paper [MMcC, footnote at p. 167], where the question of its actual occurrence in hydrodynamics is raised.

†Institute of Mathematics of the Romanian Academy P.O. Box 1-764, Ro-7070, Bucharest, ROMANIA

‡This paper was written in final form while the author was visiting the Institute for Mathematics and Its Application and Army High Performance Computing Research Center (Army Research Office contract number DAALO3-89-C-0038) whose support I would like to acknowledge.

1. Characterization and representation of injective operators in the 3-dimensional Hilbert space. In what follows H will be a real Hilbert space of dimension 3. We consider in H differential equations of the type:

$$\dot{u} + Au + B(u) = f \tag{1.1}$$

where $f \in H$ and

H1. A is a linear self-adjoint and positive operator in H;

H2. $B : H \to H$ is a homogeneous polynomial of second degree with the orthogonality property

$$\langle B(u), u \rangle = 0, \qquad \forall u \in H. \tag{1.2}$$

The hypotheses on A and B represent the essential algebraic and geometric properties of the generator of the Navier-Stokes dynamical system in a certain infinite dimensional Hilbert space. In that sense the equation (1.1) can be considered as a finite dimensional model for the Navier-Stokes system.

The condition $dim H = 3$ is linked to the fact that only for this dimension a canonical form for the nonlinear term B is available.

More precisely, we have the following

PROPOSITION 1.1. *The formula*

$$B(u) = u \times Mu, \quad u \in H, \tag{1.3}$$

gives an isomorphism between the polynomials B satisfying (H2) in the Hilbert space of dimension 3 and the zero trace operators M in the oriented space.

Proof. The main task is to find for each polynomial B with the orthogonality property (1.2) a bilinear mapping $C : H \times H \to H$ with

$$\langle C(u,v), w \rangle = -\langle C(u,w), v \rangle, \quad \forall u, v, w \in H, \tag{1.4}$$

such that

$$B(u) = C(u, u), \quad \forall u \in H. \tag{1.5}$$

By the Lemma from [M2], the map $C \mapsto B$ defined by (1.5) applies indeed the space of bilinear mappings C with (1.4) onto the space of polynomials B with (H2) and moreover is an isomorphism when restricted to the subspace of $C - s$ satisfying

$$C(u, \cdot) = C(\cdot, u) - C(\cdot, u)^*, \qquad \forall u \in H; \tag{1.6}$$

then the inverse map $B \longmapsto \tilde{B}$ is defined by

$$(1.7) \qquad \tilde{B}(u, \cdot) = \frac{1}{3}[B''(0)(u, \cdot) - B''(0)(u, \cdot)^*], \quad u \in H.$$

This fact is true for each real Hilbert space of finite dimension.

Let us remind now that in the oriented Hilbert space of dimension 3, where each monomial $u \wedge v \wedge w$ has a sign, the scalar product defines the vector product x by the condition that $\langle u \times v, w \rangle$ equals, for positive $u \wedge v \wedge w$, the volume of the parallelepiped with edges determined by u, v, w. Moreover, the mapping

$$H \to L_a(H)$$
$$v \longmapsto \cdot \times v$$

establishes an isomorphism between H and the space $L_a(H)$ of antisymmetric operators in H. Consequently the formula

$$(1.8) \qquad C(u, v) = v \times Mu, \quad u, v \in H,$$

gives an isomorphism between the space of the mappings C with (1.4) and the space of linear operators M in H. Because of the general identity in a vector space of dimension n:

$$(Mv_1) \wedge v_2 \wedge \cdots \wedge v_n + v_1 \wedge (Mv_2) \wedge \cdots \wedge v_n + \cdots + v_1 \wedge v_2 \wedge \cdots \wedge (Mv_n)$$
$$= (trM) \cdot v_1 \wedge v_2 \wedge \cdots \wedge v_n$$

the isomorphism defined by (1.8) applies the subspace of the mappings C with (1.6) onto the subspace of the operators M of zero trace.

The proof is complete. □

We are looking now for a description of the injective operators of type $A + B$ submitted to (H1) and (H2), having in view also the stability problem for the unique steady-state equilibrium of the dynamical system (1.1) they generate.

In order to prove the results we are aiming for we need the following

LEMMA 1.2. *For a linear operator M in the oriented Hilbert space H of dimension 3 the following assertions are equivalent:*

(i) $\exists u_o \neq 0$ *in H such that* $\langle u \times Mu, u_o \rangle = 0, \quad \forall u \in H;$

(ii) $det(u \times M \cdot + \cdot \times Mu) = 0, \quad \forall u \in H;$

(iii)   *the minimal polynomial of M is of degree ≤ 2;*

(iv)   *M has an eigenvalue of geometrical multiplicity ≥ 2.*

Proof. The equivalence (iii) $\Longleftrightarrow$ (iv) is general in dimension 3; we point out only that from (iii) or (iv) it follows equally that the spectrum of M is real.

On the other hand (ii) is equivalent to

(ii') $\qquad\qquad \forall u \in H \qquad \exists v \neq 0 \quad \text{in} \quad H \quad \text{such that} \quad u \times Mv + v \times Mu = 0$

while a nontrivial, but well known equivalence holds between (iii) and

(iii') $\qquad\qquad \forall u \in H \qquad \exists P_u \in \mathbf{R}[X] \setminus \{0\}, deg P_u \leq 2 \quad \text{such that} \quad P_u(M)u = 0.$

We prove the equivalence $(ii') \Longleftrightarrow (iii')$. For a vector $u \in H$ with $u \times Mu \neq 0$ the property (ii') ensures the existence of $v \neq 0$ and $\kappa_1 \in \mathbf{R}$ verifying

$$(1.9) \qquad\qquad u \times Mv = (Mu) \times v = \kappa_1 \cdot u \times Mu,$$

because $u \times Mu$ generates, in that case, the orthogonal complement of the space spanned by u and Mu. Obviously (1.9) holds also in the case when $u \times Mu = 0, u \neq 0$, for $v = u$ and $\kappa_1 = 1$.

Therefore for each $u \neq 0$ there exist $v \neq 0$ and $\kappa_o, \kappa_1, \kappa_2 \in \mathbf{R}$ such that

$$(1.10) \qquad\qquad v = -\kappa_1 \cdot u + \kappa_2 \cdot Mu,$$
$$(1.11) \qquad\qquad Mv = \kappa_1 \cdot Mu + \kappa_o \cdot u,$$

with $\kappa_2 \neq 0$ in the case that $u \times Mu \neq 0$ and $\kappa_1 \neq 0$ for $u \times Mu = 0$. From (1.10) and (1.11) we get

$$(1.12) \qquad\qquad \kappa_2 M^2 u - 2\kappa_1 Mu - \kappa_o u = 0,$$

hence a nonzero polynomial of degree ≤ 2 in M vanishing on u.

Conversely, taking for $u \neq 0$ the nonzero polynomial $P_u \in \mathbf{R}[X]$ of minimum degree such that $P_u(M)u = 0$, hence of degree ≤ 2 by the hypothesis (iii'), from (1.12) we infer that v defined by (1.10) is nonzero and satisfies (1.11), whence it follows (1.9).

Thus (ii), (iii) and (iv) are equivalent.

The inference (i) $\Longrightarrow$ (ii) is immediate, $u_o \neq 0$ being orthogonal on the range of each operator $u \times M \cdot + v \cdot \times Mu$. We show that (iv) $\Longrightarrow$ (i). If M is a scalar operator, $u_o \neq 0$ may be taken arbitrary; otherwise, there exists $\lambda \in \mathbf{R}$ for which $dim ker(M - \lambda I) = 2$ and then $dim R(-\lambda I) = 1$. Since for $u_o \neq 0, u_o \in R(M - \lambda I)$ we have $R(M - \lambda I) = \mathbf{R} \cdot u_o$, it follows that

$$\langle u \times Mu, u_o \rangle = \langle u \times (M - \lambda I)u, u_o \rangle = 0.$$

This ends the proof of the lemma. $\square$

The next theorem contains the desired characterization of the injective operators in dimension 3.

THEOREM 1.3. *In the Hilbert space H of dimension 3 an operator of type $A+B$ with (H1) and (H2), $B \neq 0$, is injective if and only if the set $\{u \in H | B(u) = 0\}$ is the union of two subspaces N_1 and N_2, $dim N_1 = 1, dim N_2 = 2, N_1 \cap N_2 = \{0\}$ and*

$$(1.13) \qquad\qquad AN_1 = N_2^{\perp}.$$

Proof. The identity

$$(1.14) \quad [Au_1 + B(u_1)] - [Au_2 + B(u_2)] = [A + B'(\frac{1}{2}(u_1 + u_2))](u_1 - u_2), \quad \forall u_1, u_2 \in H,$$

shows that the operator $A + B$ is injective if and only if

$$(1.15) \qquad\qquad det(A + B'(u)) \neq 0, \quad \forall u \in H.$$

In order to ensure (1.15) the homogeneous part of degree 3 in the polynomial $det(A + B'(u))$ must vanish, that is

$$(1.16) \qquad\qquad det(B'(u)) = 0, \quad \forall u \in H.$$

For the operator M defining B by (1.3) this is precisely the property (ii) of the Lemma 1.2. From the equivalent to it property (iv) and the fact that $B \neq 0$ we infer the existence of a unique operator M_o in H such that

$$(1.17) \qquad \begin{cases} B(u) = u \times M_o u, & \forall u \in H, \\ dim\ ker M_o = 2. \end{cases}$$

Let us denote

$$(1.18) \qquad\qquad N_1 = R(M_o), \qquad N_2 = ker M_o$$

and distinguish the cases

$$(\alpha) \quad N_1 \not\subset N_2 \quad (\text{i.e.} \quad N_1 \cap N_2 = \{0\});$$
$$(\beta) \quad N_1 \subset N_2.$$

In the case (α) M_o is diagonalizable and N_1 is an eigenspace; in the case $\{\beta\}$ M_o is nilpotent of order two. Anyway, we can choose the unitary vectors e_1, e_2, e_3 such that

$$(1.19) \qquad e_3 \in N_2^{\perp}, e_2 \in (N_2^{\perp} + N_1)^{\perp}, e_1 \in \{e_2, e_3\}^{\perp}.$$

and $e_1 \wedge e_2 \wedge e_3$ positive.

Thus we get an orthonormal basis in H with respect to which M_o is represented by a matrix of the form

$$(1.20) \qquad M_o = \begin{pmatrix} 0 & 0 & \sigma \\ 0 & 0 & 0 \\ 0 & 0 & \tau \end{pmatrix}, with \quad \sigma, \tau \in \mathbf{R}$$

The case (α) corresponds then to $\tau \neq 0$ and the case (β) to $\tau = 0$ and $\sigma \neq 0$.

Accordingly, the operator $B'(u) = u \times M_o \cdot + \cdot \times M_o u$ is represented by the matrix

$$(1.21) \qquad B'(u) = \begin{pmatrix} 0 & \tau u_3 & \tau u_2 \\ -\tau u_3 & 0 & 2\sigma u_3 - \tau u_1 \\ 0 & -\sigma u_3 & -\sigma u_2 \end{pmatrix}.$$

Consequently the polynomial $det(A + B'(u))$ will contain only the monomials $u_3^2, u_1 u_3, u_2 u_3, u_1, u_2, u_3$ and the constant term. Therefore (1.15) holds only if the coefficients of $u_1 u_3, u_2 u_3, u_1,$ and u_2 vanish. If a_{ij} denote the matrix coefficients of A in the basis (1.19), these conditions read respectively

$$(1.22) \qquad \tau(\tau a_{31} + \sigma a_{11}) = 0,$$

$$(1.23) \qquad \tau(\tau a_{32} + \sigma a_{12}) = 0,$$

$$(1.24) \qquad \tau(a_{11} a_{32} - a_{12} a_{31}) = 0,$$

$$(1.25) \qquad \tau(a_{21} a_{32} - a_{22} a_{31}) - \sigma(a_{11} a_{22} - a_{12} a_{21}) = 0.$$

As $a_{11} a_{22} - a_{12} a_{21} \neq 0$, A being definite, it is clear that $\tau = 0$ entails by (1.25) $\sigma = 0$. Thus the case (β) is excluded: we have surely $\tau \neq 0$. From (1.22) and (1.23) we infer now

$$(1.26) \qquad \tau a_{31} + \sigma a_{11} = 0,$$

$$(1.27) \qquad \tau a_{32} + \sigma a_{12} = 0,$$

and remark that (1.24) and (1.25) are consequences of these two relations. They can also be written in the form $\langle Ae_1, M_o e_3 \rangle = 0$, $\langle Ae_2, M_o e_3 \rangle = 0$, giving thus precisely

$$(1.28) \qquad AN_2 = N_1^{\perp}.$$

Conversely, suppose that A and B fulfill the conditions of the theorem. Then the basis (1.19) is correctly defined and in this basis (see (1.26), (1.27))

$$(1.29) \qquad det(A + B'(u)) = [a_{33} - (\frac{\sigma}{\tau})^2 a_{11}] \cdot (\tau^2 u_3^2 + a_{11} a_{22} - a_{12}^2).$$

Hence the condition (1.13) (or the equivalent to it (1.28)) is also sufficient for the injectivity of the operator $A + B$. The theorem is proved. $\square$

Remark 1.4. Let $A + B$ be an injective operator with $B \neq 0$ as in the theorem above. Then $\tau \neq 0$ and denoting (see (1.20))

$$(1.30) \qquad \xi = -\frac{\sigma}{\tau}$$

we derive from (1.21), (1.26) and (1.27) the following expressions for its linear and nonlinear part in the basis (1.19):

$$(1.31) \qquad A = \begin{pmatrix} a_{11} & a_{12} & \xi a_{11} \\ a_{12} & a_{22} & \xi a_{12} \\ \xi a_{11} & \xi a_{12} & a_{33} \end{pmatrix}$$

$$(1.32) \qquad B(u) = \tau \cdot \begin{pmatrix} u_2 u_3 \\ -\xi u_3^2 - u_1 u_3 \\ \xi u_2 u_3 \end{pmatrix}.$$

Taking into account the Sylvester-Jacobi criterion and the formula (see(1.29))

$$det A = (a_{33} - \xi^2 a_{11})(a_{11} a_{22} - a_{12}^2)$$

we see that the matrix of A is positively definite if and only if

$$(1.33) \qquad a_{11} > 0, \quad a_{11} a_{22} - a_{12}^2 > 0, \quad a_{33} - \xi^2 a_{11} > 0.$$

These inequalities will be assumed to hold and used throughout from now on.

The expressions (1.31) and (1.32) define an injective operator for all $\tau \in \mathbf{R}$ and all a_{ij}, ξ submitted to (1.33), such that $B = 0$ if and only if $\tau = 0$. They show that in the case $B \neq 0$ we may suppose $\tau = 1$ by rescaling the phase space: $u \longmapsto \tau u$.

Remark 1.5. For an injective operator with $B \neq 0$ the number ξ has a geometrical meaning:

$$(1.34) \qquad |\xi| = |cotg(\widehat{N_1, N_2})|.$$

Hence $\xi = 0$ if and only if $N_1 \perp N_2$.

Remark 1.6. We will call *generic injective* an injective operator $A + B$ with $\tau \neq 0$ and $\xi \neq 0$, i.e. $B \neq 0$ and $N_1 \neq N_2^\perp$. For such an operator e_2 has the direction of $\nabla_u tr(A + B'(u))$ because [see (1.32)]

$$(1.35) \qquad tr[A + B'(u)] = tr A + \tau \xi u_2.$$

This direction is the natural one in conjunction with the Hopf bifurcation as will be seen later on. On the other hand e_3 has the direction of $\nabla_u det(A + B'(u))$ in the points where the gradient is nonzero:

$$(1.36) \qquad det(A + B'(u)) = (a_{33} - \xi^2 a_{11})(\tau^2 u_3^2 + a_{11} a_{22} - a_{12}^2)$$

[see (1.29)] □

We are now interested to know the supplementary conditions under which the unique stationary solution of the equation (1.1) is stable for all $f \in H$(the example given in [M1] being of that kind). The answer is surprisingly simple.

THEOREM 1.7. *Under the conditions and with the notations of the preceding theorem the spectrum of the linearized operator $A + B'(u)$ of an injective operator $A + B, B \neq 0$, remains in the right open half plane for all $u \in H$ if and only if $N_1 = N_2^\perp$.*

Proof. From (1.35) we see that the spectrum of $A + B'(u)$ may remain in the right open half plane for all $u \in H$ only if $\xi = 0$.

This condition is also sufficient. Indeed, for $\xi = 0$ we have according to (1.31) and (1.32)

(1.37) $\quad det(xI - A - B'(u)) = (x - a_{33})[x^2 - (a_{11} + a_{22})x + a_{11}a_{22} - a_{12}^2 + \tau^2 u_3^2],$

which obviously has the roots in the right open half plane. The proof is complete. ▯

For the 2D-Navier-Stokes equations on a compact surface without boundary (in particular, for the periodic in space equations) the following supplementary orthogonality holds (see[T], [CF])

(1.38) $\qquad\qquad\qquad \langle B(u), Au \rangle = 0, \quad \forall u \in H.$

To what extent is substantial the axiomatics (H1), (H2) can be seen also from the following result which answers in the affirmative a question raised by C. Foias.

THEOREM 1.8. *If the injective operator $A + B, B \neq 0$ in the Hilbert space of dimension 3 satisfies also (1.38) then $N_1 = N_2^{\perp}$. Moreover $A|N_2$ is a scalar operator.*

Proof. The assertions can be proved straightforwardly by identifying the coefficients in the identity (1.38). Yet we like better another justification of the first one. The operator $A^2 + AB$ verifies the hypotheses (H1) and (H2) and because $A^2 + AB = A(A + B)$ it is also injective. On the other hand $AB(u) = 0$ if and only if $B(u) = 0$. In accordance with the theorem 1.3

$$AN_1 = N_2^{\perp} \quad \text{and} \quad A^2 N_1 = N_2^{\perp} \quad \text{too,}$$

whence $AN_1 = N_1$ and then $N_1 = N_2^{\perp}$ ▯

In the case of the Navier-Stokes equations there exists a Lie algebra structure which defines the non-linear term by the formula (see[A, Appendix 2]).

(1.39) $\qquad\qquad\qquad \langle B(u), v \rangle = \langle u, [v, u] \rangle, \qquad \forall u, v \in H.$

We show now that a generic injective operator (see the Remark 1.6) can be realized in the same manner on a certain Lie algebra endowed with a scalar product.

THEOREM 1.9. *Let $A + B$ be a generic injective operator in H, $dim H = 3$. Then there exists a unique Lie algebra structure on H such that the quadratic term be defined by (1.39); this structure is always isomorphic to the Lie algebra of matrices $\begin{pmatrix} a & c \\ 0 & b \end{pmatrix}$.*

Proof. Let $D : H \times H \to H$ be an arbitrary bilinear mapping with the properties

(1.40) $\qquad D(u, v) = -D(v, u), \quad \langle B(u), v \rangle = \langle u, D(v, u) \rangle, \qquad \forall u, v \in H.$

It follows that the trilinear functional on H

$$\langle M_o w, v \times u \rangle - \langle w, D(v, u) \rangle$$

is alternate and hence there exists $c \in \mathbf{R}$ such that

$$\langle M_o w, v \times u \rangle - \langle w, D(v, u) \rangle = c \langle w, v \times u \rangle,$$

or equivalently

$$D(u, v) = (M_o - cI)^*(u \times v), \qquad \forall u, v \in H.$$

According to [GG, 1.7] a mapping of the form $D(u, v) = T(u \times v)$ verifies the Jacobi identity if and only if the transposed of the matrix of algebraic complements of T written in an orthonormal basis is symmetric. The matrix of $(M_o - cI)^*$ in the basis (1.19) being

$$\begin{pmatrix} -c & 0 & 0 \\ 0 & -c & 0 \\ -\tau \xi & 0 & \tau - c \end{pmatrix}$$

the transposed of the matrix of algebraic complements will be of the form

$$\begin{pmatrix} * & 0 & 0 \\ 0 & * & 0 \\ -c\tau \xi & 0 & * \end{pmatrix}.$$

As $\tau \xi \neq 0$ this matrix is symmetric only for $c = 0$. Hence there exists a unique Lie algebra structure on H verifying (1.39) and its bracket is defined by

(1.41) $$[u, v] = M_o^*(u \times v).$$

It is easy to see that for this structure we have

(1.42) $$[H, H] = N_2^\perp, \qquad Z(H) = N_1.$$

On the other hand each Lie algebra of dimension 3 with the properties $dim[H, H] = 1$ and $[H, H] \not\subset Z(H)$ is isomorphic to the Lie algebra of matrices $\begin{pmatrix} a & c \\ 0 & b \end{pmatrix}$ (see, for instance, [GG, 1.7]). This ends the proof of the theorem. $\square$

Remark 1.10. A generic injective operator $A + B$ is thus realized by a scalar product on the Lie algebra h of the matrices $\begin{pmatrix} a & c \\ 0 & b \end{pmatrix}$ with the property

(1.43) $$Z(h) \not\subset [h, h].$$

and a positive self-adjoint with respect to it operator A with

(1.44) $$AZ(h) = [h, h].$$

(B being defined by (1.39))

Remark 1.11. In the case that the Hilbert space H enjoys a Lie algebra structure defining the non-linear term B by (1.39), the Lie bracket provides another mapping C with (1.4) and (1.5), namely

$$C_o(u, v) = -(ad_v)^* u,$$

where $ad_v w = [v, w]$. In the case of a generic injective operator, M_o corresponds to C_o defined by the canonic Lie algebra structure.

2. The surface of change of stability. In what follows $A + B$ will be a fixed generic injective operator submitted to (H1) and (H2) in the Hilbert space H of dimension 3. Taking into account the Remark 1.4 we consider $\tau = 1$ in (1.32); we may suppose also $\xi > 0$ by the eventual change of basis $e_1 \mapsto -e_1, e_2 \mapsto -e_2, e_3 \mapsto e_3$ (see (1.19), (1.20) and (1.30)).

$f \in H$ will be seen as variable parameter in (1.1) and because of the one-to-one correspondence the unique stationary solution of (1.1) can be taken as parameter of bifurcation as we will do in the sequel.

Let us denote

$$(2.1) \qquad D(u) = A + B'(u), \qquad u \in H$$

and remind that

$$(2.2) \qquad det D(u) > 0, \qquad \forall u \in H.$$

We state without proof the following

PROPOSITION 2.1. *Let H be a real vector space of dimension 3 and $D \in L(H)$ a linear operator with*

$$(2.3) \qquad det D > 0.$$

Then

(i) the spectrum of D lies in the right open half plane if and only if

$$(2.4) \qquad det(tr D \cdot I - D) > 0 \quad and \quad tr D > 0;$$

(ii) D has two conjugate eigenvalues on the imaginary axis if and only if

$$(2.5) \qquad det(tr D \cdot I - D) = 0 \quad and \quad tr D > 0.$$

Remark 2.2. The case

$$(2.6) \qquad det(tr D \cdot I - D) = 0 \quad and \quad tr D < 0$$

corresponds to a pair of opposite real eigenvalues.

Accordingly, if the stationary solution $u \in H$ of (1.1) is taken as parameter then the open set of stability is given by

$$(2.7) \qquad \begin{cases} det(tr D(u) \cdot I - D(u)) > 0, \\ tr D(u) > 0 \end{cases}$$

and the the surface of change of stability by

$$(2.8) \qquad \begin{cases} det(tr D(u) \cdot I - D(u)) = 0, \\ tr D(u) > 0. \end{cases}$$

Let us denote

$$(2.9) \qquad \sum = \{u \in H | det(tr D(u) \cdot I - D(u)) = 0\},$$

$$(2.10) \qquad \prod = \{u \in H | tr D(u) = 0\},$$

$$(2.11) \qquad \sum{}_+ = \sum \cap \{u | tr D(u) > 0\},$$

$$(2.12) \qquad \sum{}_- = \sum \cap \{u | tr D(u) < 0\}.$$

THEOREM 2.3. *The plane $\prod$ does not meet $\sum$ and separates $\sum_+$ of $\sum_-$. $\sum_+$ and $\sum_-$ are smooth algebraic manifolds intersected by each straight line orthogonal to $\prod$ transversally in precisely one point.*

Proof. From (1.21) and (1.31) we derive successively

$$(2.13) \qquad D(u) = \begin{pmatrix} a_{11} & a_{12} + u_3 & \xi a_{11} + u_2 \\ a_{12} - u_3 & a_{22} & \xi a_{12} - 2\xi u_3 - u_1 \\ \xi a_{11} & \xi(a_{12} + u_3) & a_{33} + \xi u_2 \end{pmatrix}$$

$$tr D(u) \cdot I - D(u) = \begin{pmatrix} a_{22} + a_{33} + \xi u_2 & -(a_{12} + u_3) & -(\xi a_{11} + u_2) \\ u_3 - a_{12} & a_{11} + a_{33} + \xi u_2 & u_1 + 2\xi u_3 - \xi a_{12} \\ -\xi a_{11} & -\xi(a_{12} + u_3) & a_{11} + a_{22} \end{pmatrix}.$$

Therefore the polynomial

$$(2.14) \qquad P(u) = det(tr D(u) \cdot I - D(u))$$

is of the form

$$(2.15) \qquad P(u_1, u_2, u_3) = \xi^2 a_{22} u_2^2 + Q(u_1, u_3) u_2 + R(u_1, u_3),$$

hence of second degree with respect to u_2 with an independent of (u_1, u_3) and positive coefficient by u_2^2. On the other hand, for all $u \in \prod$, $P(u) = -det D(u) < 0$. Thus $\sum_+$ and $\sum_-$ are respectively the graphs of the roots of $P(u_1, \cdot, u_3)$ as functions of (u_1, u_3) □

3. On the existence of a global Lyapunov functional. Here $A + B$ will be again a fixed injective operator with $B \neq 0$. We consider then as before $\tau = 1$ and $\xi \geq 0$.

Being interested in the evolution of the attraction basin of the stationary solution u we look first for those values of u as parameter corresponding to a globally attracting steady-state equilibrium. A sufficient condition for that is the existence of a global Lyapunov functional for the stationary solution u.

We call *global Lyapunov functional for u* a positive definite quadratic form l on H such that

$$(3.1) \qquad \frac{d}{dt} l(u(t) - u) < 0,$$

for all $t \in \mathbf{R}$ and all solution $u(t)$, different from u, of (1.1).

Let Λ be the positive self-adjoint operator in H defining l by

$$(3.2) \qquad l(v) = \langle \Lambda v, v \rangle.$$

Then the following characterization holds:

PROPOSITION 3.1. Λ defines by (3.2) a global Lyapunov functional for u if and only if

$$(3.3) \qquad\qquad \Lambda D(u) + D(u)^*\Lambda > 0$$

and

$$(3.4) \qquad\qquad \langle B(v), \Lambda v \rangle = 0, \qquad \forall v \in H$$

(The inequality (3.3) has to be understood in the sense of self-adjoint operators in H, $D(u)$ being the linearization of $A + B$ at u).

Proof. If $u(t)$ is an arbitrary solution of (1.1) then $v(t) = u(t) - u$ verifies the equation

$$(3.5) \qquad\qquad Dv(t) + D(u)v(t) + B(v(t)) = 0.$$

Taking the scalar product of it by $\Lambda v(t)$, we see that (3.1) is equivalent to

$$\langle \Lambda D(u)v, v \rangle + \langle B(v), \Lambda v \rangle > 0, \qquad \forall v \in H, \quad v \neq 0.$$

Replacing here v by tv, with $t \in \mathbf{R}\backslash\{0\}$, we see that this inequality can be fulfilled only if

$$\langle \Lambda D(u)v, v \rangle > 0 \quad \text{and} \quad \langle B(v), \Lambda v \rangle = 0, \qquad \forall v \in H, \quad v \neq 0.$$

☐

From this result we derive as usual

THEOREM 3.2. If the stationary solution u admits a global Lyapunov functional then it is globally attracting; more precisely, for each $\mu > 0$ such that

$$\Lambda D(u) + D(u)^*\Lambda \geq \mu\Lambda$$

and each solution $u(t)$ of (1.1)

$$l(u(t) - u) \leq e^{-\mu t}l(u(0) - u), \qquad \forall t \geq 0. \quad \square$$

In order to find the values of the parameter f for which the corresponding stationary solution admits global Lyapunov functional we need first the following:

PROPOSITION 3.3. Let $A + B$ injective with $B \neq 0$. A positive self-adjoint operator Λ verifies (3.4) if and only if it is of the form

$$(3.6) \qquad\qquad \Lambda = \lambda P_1 + \mu(I - P_1), \qquad \lambda, \mu > 0$$

where P_1 denotes the orthogonal projection on N_1.

Proof. As ΛB too enjoys the orthogonality property there exists N linear operator such that

$$(3.7) \qquad \Lambda(u \times Mu) = u \times Nu, \qquad \forall u \in H.$$

Each eigenvector of M is then also an eigenvector for N whence it follows that N_1 and N_2 are eigenspaces for N too. Therefore there exist $\mu, \nu \in \mathbf{R}$ such that

$$N = \mu M + \nu I;$$

this relation combined with (3.7) gives

$$(3.8) \qquad \Lambda(u \times Mu) = \mu(u \times Mu), \qquad \forall u \in H,$$

which shows that the linear span $sp\{B(u)|u \in H\}$ is an eigenspace for Λ. From the equalities (see (1.18))

$$B(u) = u \times M_o u = u_3(u \times M_o e_3)$$

it follows readily that

$$(3.9) \qquad sp\{B(u)|u \in H\}^\perp = N_1.$$

Thus $N_1^\perp$ and N_1 are eigenspaces for Λ. The proof is complete. □

Remark 3.4. This result together with the Theorem 1.3 entails immediately the Theorem 1.8.

We are now ready to prove the central result of this paragraph.

THEOREM 3.5. *Let $A + B$ injective, $\tau = 1$. The unique stationary solution of (1.1) admits global Lyapunov functional if and only if, in the basis (1.19)*

$$(3.10) \quad [(\xi^2 + 1)a_{12} - \frac{\xi}{2}(\xi u_3 + u_1)]^2 < a_{22}[\xi(\xi^2 + 1)u_2 + (2\xi^2 + 1)a_{11} + \xi^2 a_{33}].$$

Proof. From (1.18), (1.20) and (1.30) we get first

$$(3.11) \qquad N_1 = \mathbf{R} \cdot \begin{pmatrix} -\xi \\ 0 \\ 1 \end{pmatrix}$$

and from (3.6) we derive next the matrix of an operator Λ with (3.4) in the basis(1.19):

$$(3.12) \qquad \Lambda = p \cdot \begin{pmatrix} q\xi^2 + 1 & 0 & \xi(1-q) \\ 0 & \xi^2 + 1 & 0 \\ \xi(1-q) & 0 & q + \xi^2 \end{pmatrix}, \qquad p, q > 0.$$

Taking $p = 1$, which is legitimate since the inequality (3.3) is independent of the positive p, we get (see (2.13))

$$(3.13) \qquad \frac{1}{2}(\Lambda D(u) + D(u)^*\Lambda) = \begin{pmatrix} b_{11} & b_{12} & \frac{1}{2}(s + \xi b_{11} - \xi x) \\ b_{12} & b_{22} & t \\ \frac{1}{2}(s + \xi b_{11} - \xi x) & t & \xi s + x \end{pmatrix},$$

where

$$(3.14) \qquad b_{ij} = (\xi^2 + 1)a_{ij}, \qquad 1 \le i, j \le 2$$

$$(3.15) \qquad s = (\xi^2 + 1)u_2 + \xi(a_{11} + a_{33})$$

$$(3.16) \qquad t = \frac{1}{2}(\xi^2 + 1)(2\xi a_{12} - \xi u_3 - u_1)$$

$$(3.17) \qquad x = q(a_{33} - \xi^2 a_{11}).$$

The problem at hand is hence to characterize those values of s and t for which the matrix (3.13) becomes positively definite for at least one positive x. Since the matrix $(b_{ij})_{i,j=1}^2$ is positively definite, according to the Sylvester-Jacobi criterion the matrix (3.13) is positively definite if and only if its determinant is positive.

As

$$(3.18)$$
$$det[\frac{1}{2}(\Lambda D(u) + D(u)^*\Lambda)] = -\frac{b_{22}\xi^2}{4} \cdot x^2 + \left[\frac{b_{22}\xi}{2}(s + \xi b_{11}) - b_{12}\xi t + b_{11}b_{22} - b_{12}^2 \right] x$$
$$+ b_{12}t(s + \xi b_{11}) - \frac{b_{22}}{4}(s + \xi b_{11})^2 - b_{11}t^2 + \xi s(b_{11}b_{22} - b_{12}^2),$$

it becomes positive for a real x if and only if the discriminant Δ with respect to x is positive. But

$$(3.19) \qquad \Delta = (b_{11}b_{22} - b_{12}^2)[b_{22}(\xi^2 + 1)(\xi s + b_{11}) - (\xi t + b_{12})^2].$$

On the other hand the sum S of the roots of this polynomial in x can be written in the following form

$$(3.20)$$
$$S = \frac{2}{b_{22}\xi^2} \Big\{ \frac{1}{\xi^2 + 1}[b_{22}(\xi^2 + 1)(\xi s + b_{11}) - (\xi t + b_{12})^2]$$
$$+ \frac{1}{\xi^2 + 1}[(\xi^2 + 1)b_{12} - (\xi t + b_{12})]^2 + (\xi^2 + 1)(b_{11}b_{22} - b_{12}^2) \Big\}$$

which shows that for positive Δ the sum S is positive too. Therefore if there exists a real x making (3.18) positive then there exists even a positive x, so that the polynomial (3.18) is positive for a positive x if and only if $\Delta > 0$. This is the condition we were looking for. It can be written according to (3.19)

$$(\xi t + b_{12})^2 < b_{22}(\xi^2 + 1)(\xi s + b_{11})$$

and using (3.14), (3.15), (3.16) in the form (3.10).

$\square$

COROLLARY 3.6. *Let* $\dim H = 3$. *If the stationary solution of the equation (1.1) is unique and stable for each* f *then it is globally attracting for each* f.

$\square$

According to the easy part of the Lyapunov Theorem the domain (3.10) is included in the open set of stability (2.7). Yet it is not obvious that its boundary

$$(3.21) \quad \Xi = \{u \in H \mid [(\xi^2 + 1)a_{12} - \frac{\xi}{2}(\xi u_3 + u_1)]^2$$
$$= a_{22}[\xi(\xi^2 + 1)u_2 + (2\xi^2 + 1)a_{11} + \xi^2 a_{33}]\}$$

is itself contained in the domain of stability. Precisely this is the content of the following theorem.

THEOREM 3.7. Ξ *is a smooth algebraic manifold intersected by each straight line orthogonal to the plane* $\prod$ *(see(2.10)) transversally in precisely one point. Moreover* Ξ *does not meet the surface of change of stability* $\sum_+$.

Proof. The first assertion is quite obvious from the equation of Ξ. For the second we write the polynomial (2.14) in the following form obtained from (2.13)

$$(3.22) \quad \begin{aligned} P(u) &= \xi(a_{12} + u_3)(u_1 + \xi u_3)(tr A + \xi u_2) \\ &+ (u_3^2 - a_{12}^2)[(\xi^2 + 1)(tr A + \xi u_2) + \xi^2 a_{11} - a_{33}] \\ &+ (tr A + \xi u_2 - a_{22})[a_{22}(tr A + \xi u_2) + a_{11}(a_{33} - \xi^2 a_{11})]. \end{aligned}$$

Then in the new coordinates

$$(3.23) \quad \left\{ \begin{aligned} x &= \tfrac{\xi}{2}(\xi u_3 + u_1) - (\xi^2 + 1)a_{12} \\ y &= (\xi^2 + 1)(tr A + \xi u_2 - a_{22}) + \xi^2 a_{11} - a_{33} \\ z &= \tfrac{\xi}{2}(\xi u_3 + u_1) + (\xi^2 + 1)u_3 \end{aligned} \right.$$

and with the notation $b = a_{33} - \xi^2 a_{11}$, the expression of P becomes

$$(3.24) \quad \begin{aligned} P(x, y, z) &= 2\frac{1}{(\xi^2 + 1)^2}(z - x)[x + (\xi^2 + 1)a_{12}][y + (\xi^2 + 1)a_{22} + b] \\ &+ \frac{1}{\xi^2 + 1}(z - x)[\frac{1}{\xi^2 + 1}(z - x) - 2a_{12}][y + (\xi^2 + 1)a_{22}] \\ &+ \frac{1}{\xi^2 + 1}(y + b)[\frac{a_{22}}{\xi^2 + 1}(y + (\xi^2 + 1)a_{22} + b) + a_{11}b] \end{aligned}$$

and the equation of Ξ is

$$(3.25) \quad x^2 = a_{22}y.$$

Now the following identity

$$(3.26) \quad \begin{aligned} P(x, \frac{x^2}{a_{22}}, z) &= \frac{(xz + a_{22}b)^2}{a_{22}(\xi^2 + 1)^2} + \frac{(a_{22}z + a_{12}b)^2}{a_{22}(\xi^2 + 1)} + \frac{b(a_{11}x - a_{12}a_{22})^2}{a_{11}a_{22}(\xi^2 + 1)} \\ &+ \frac{b(a_{11}a_{22} - a_{12}^2)(a_{22}^2 + a_{11}b)}{a_{11}a_{22}(\xi^2 + 1)} \end{aligned}$$

shows that P is positive in the points of Ξ. $\square$

4. The Hopf Bifurcation. Let as before $A + B$ be a generic injective operator in H of dimension 3 and let

$$(4.1) \qquad \sigma(D(u)) = \{\lambda(u), \overline{\lambda(u)}, \mu(u)\}$$

be the spectrum of $D(u)$ in a neighborhood of $\sum_+$. If a path $u(s)$ in H is crossing $\sum_+$ for the value $s = 0$ then the pair of eigenvalues $\lambda(u(s)), \overline{\lambda(u(s))}$ will be crossing the imaginary axis in the complex plane. The sufficient condition for the bifurcation of a locally unique, around the point $(0,0)$, one parameter family of nontrivial closed orbits of the vector field in $H \times \mathbf{R}$

$$(4.2) \qquad X(v, s) = (-D(u(s))v - B(v), 0)$$

from the line of stationary points $\{(0, s) | s \in \mathbf{R}\}$, is that

$$(4.3) \qquad \frac{d}{ds} Re\,\lambda(u(s))|_{s=0} \neq 0$$

(see [MMcC, sections 3 and 5]).

In our case <u>this relation holds as soon as</u> $u(s)$ is transversal in $s = 0$ to $\sum_+$, i.e., $\dot{u}(0) \notin T_{u(0)} \sum_+$, and <u>this is</u> in turn ensured by the property

$$(4.4) \qquad P'(u_0) \neq 0, \qquad \forall u_0 \in \sum_+.$$

the proof of the Theorem 2.3 shows more precisely that

$$(4.5) \qquad \frac{\partial P}{\partial u_2}(u_0) \neq 0, \qquad \forall u_0 \in \sum_+.$$

Indeed, in the neighborhood of $\sum_+$ we have

$$P(u) = 2Re\,\lambda(u)[(Re\,\lambda(u) + \mu(u))^2 + (Im\,\lambda(u))^2]$$

and hence for $u_0 \in \sum_+$, where $(Im\,\lambda(u_0))^2 > 0$,

$$(4.6) \qquad Ker(Re\,\lambda)'(u_0) = Ker\,P'(u_0) = T_{u_0} \sum_+.$$

Thus, if $\dot{u}(0) \notin T_{u_0} \sum_+$

$$\frac{d}{ds} Re\,\lambda(u(s))|_{s=0} = (Re\,\lambda)'(u(0)) \cdot \dot{u}(0) \neq 0.$$

In this way we have proved the quite natural

THEOREM 4.1. *For the equation (1.1), with $A + B$ generic injective and $f \in H$ parameter, the Hopf bifurcation occurs along any path $f(s)$ in H for which the corresponding path of stationary solutions $u(s)$ is crossing transversally the surface of change of stability $\sum_+$, wherever the crossing point would be. More precisely, it occurs along the paths of stationary solutions*

(4.7) $$u(s) = u_o + se_2,$$

defined by arbitrary $u_o \in \sum_+$. □

We are next concerned with the stability of the bifurcating closed orbits, being at the same time interested by the dependence of the stability on the crossing point $u_0 \in \sum_+$. Though only the stable periodic solutions, that bifurcate after the stationary solution lost the stability, are observable as such, the appearance of unstable closed orbits before the loss of stability – the subcritical bifurcation – allows to record that the attraction basin is actually shrinking when the stable stationary solution is approaching a certain region of the surface $\sum_+$.

Because of its complexity, we were not able so far to perform the calculus of the stability coefficient at a general point $u_o \in \sum_+$. We succeeded however in computing it at an arbitrary point of a certain line lying on $\sum_+$ due to a convenient spectral property which is worth stating for itself.

PROPOSITION 4.2. *For $A + B$ generic injective and $f \in H$ parameter, the mapping*

(4.8) $$\begin{cases} u \longmapsto (|\lambda(u)|, -\mu(u)) \\ \sum_+ \longrightarrow \mathbf{R}^2, \end{cases}$$

assigning to a stationary solution from the surface of change of stability its characteristic exponents, has only one infinite level set and that includes the line

(4.9) $$\delta = \{u \in H | trA + \xi u_2 = a_{22}, \quad \alpha_{12} + u_3 = 0\}.$$

Moreover, for all $u \in \delta, e_2$ is an eigenvector of $D(u)$.

Proof. The characteristic exponents determine $trD(u)$ and $detD(u)$, that is u_2 and u_3^2 respectively (see (1.35) and (1.36)). According to (3.22), for u_2 and u_3 given such that $(a_{12} + u_3)(trA + \xi u_2) \neq 0$ there is precisely one point u on $\sum$ with these components. Hence an infinite level set has to contain the points $u \in H$ satisfying

(4.10) $$\begin{cases} (a_{12} + u_3)(trA + \xi u_2) = 0, \\ (trA + \xi u_2 - a_{22})[a_{22}(trA + \xi u_2) + a_{11}(a_{33} - \xi^2 a_{11})] = 0, \\ trA + \xi u_2 > 0; \end{cases}$$

since $a_{22} > 0$, the set of points with (4.10) coincides with the straight line δ.

For $u \in \delta$ we have

(4.11) $$|\lambda(u)| = \sqrt{a_{11}(a_{33} - \xi^2 a_{11})}, \qquad \mu(u) = a_{22}.$$

(Remark that for $a_{12} \neq 0$ the level set of (4.8) corresponding to this value contains also the point of $\sum_+$ with $u_3 = a_{12}, trA + \xi u_2 = a_{22}$).

The second assertion follows immediately from (2.13). The proof is complete. □

Our result about the stability of the bifurcating closed orbit is then the following

THEOREM 4.3. *In the conditions of the Theorem 4.1 the Hopf bifurcation is supercritical and stable, wherever the crossing point would be on δ outside a bounded interval.*

In the case that

$$(4.12) \qquad 2a_{11}(\lambda^2 + a_{22}^2)[(\xi^2+1)a_{11}(3a_{22}^2 + 8\lambda^2) - 2a_{22}\lambda^2] < a_{12}^2(a_{22}^2 + 6\lambda^2)^2,$$

where

$$(4.13) \qquad \lambda = \sqrt{a_{11}(a_{33} - \xi^2 a_{11})},$$

there are points on δ around which the bifurcation is subcritical and unstable.

Proof. In computing the stability coefficient $V'''(0)$ we follow the algorithm from [MMcC, section 4A]. This coefficient depends only on the vector field at the bifurcation value $s = 0$ of the parameter s, hence on the vector field (see (4.2)).

$$(4.14) \qquad X_0(v) = -D(u_0)v - B(v),$$

$u_0 \in \sum_+$ being the crossing point; we take $u_0 \in \delta$ (see(4.9)) arbitrary but fixed. So the linearization of X_0 at the stationary point $0 \in H$ is

$$(4.15) \qquad dX_0(0) = -D(u_0).$$

According to (4.11) and (4.13)

$$|\lambda(u_0)| = \lambda, \qquad \mu(u_0) = a_{22}$$

and the first task is to choose new linear coordinates in H bringing $dX_0(0)$ to the canonical form

$$(4.16) \qquad \begin{pmatrix} 0 & \lambda & 0 \\ -\lambda & 0 & 0 \\ 0 & 0 & -a_{22} \end{pmatrix}.$$

Since $u_0 \in \delta$ has the components

$$(4.17) \qquad u_0 = \left(u_1, -\frac{1}{\xi}(a_{11} + a_{33}), -a_{12}\right)$$

we have according to (2.13)

$$(4.18) \qquad D(u_0) = \begin{pmatrix} a_{11} & 0 & \xi a_{11} - \frac{1}{\xi}(a_{11} + a_{33}) \\ 2a_{12} & a_{22} & 3\xi a_{12} - u_1 \\ \xi a_{11} & 0 & -a_{11} \end{pmatrix}.$$

As the vectors

$$v_1 = a_{11} \begin{pmatrix} a_{22} + a_{33} - \xi^2 a_{11} \\ \xi u_1 - (3\xi^2 + 2)a_{12} \\ \xi a_{22} \end{pmatrix}, \quad v_2 = \lambda \begin{pmatrix} a_{11} - a_{22} \\ 2a_{12} \\ \xi a_{11} \end{pmatrix}$$

verify the relations

$$D(u_0)v_1 = \lambda v_2, \qquad D(u_0)v_2 = -\lambda v_1,$$

the matrix

(4.19)
$$T = \begin{pmatrix} a_{11}a_{22} + \lambda^2 & \lambda(a_{11} - a_{22}) & 0 \\ x & 2\lambda a_{12} & 1 \\ \xi a_{11}a_{22} & \lambda\xi a_{11} & 0 \end{pmatrix},$$

with the notation

(4.20)
$$x = a_{11}[\xi u_1 - (3\xi^2 + 2)a_{12}],$$

makes the desired change of coordinates:

$$-T^{-1}D(u_0)T = \begin{pmatrix} 0 & \lambda & 0 \\ -\lambda & 0 & 0 \\ 0 & 0 & -a_{22} \end{pmatrix}.$$

The nonlinear part of X_0 (see (4.14) and (1.32)) can be written in the form

(4.21)
$$-B(v) = v_3 \cdot Qv,$$

with

(4.22)
$$Q = \begin{pmatrix} 0 & -1 & 0 \\ 1 & 0 & \xi \\ 0 & -\xi & 0 \end{pmatrix}.$$

Hence if we denote

(4.23)
$$R = T^{-1}QT,$$

the vector field X_0 has in the new coordinates $w = T^{-1}v$ the expression

(4.24)
$$X_0(w) = Lw + (Tw)_3 \cdot Rw,$$

where L denotes the matrix (4.16). Thus we have

(4.25)
$$\partial_i\partial_j X_0^k(0) = t_{3i}r_{kj} + t_{3j}r_{ki}, \qquad 1 \le i, j, k \le 3,$$

$t_{\alpha\beta}r_{\alpha\beta}$ denoting the matrix components of T and R respectively.

The stability coefficient has the expression

(4.26)
$$V'''(0) = S_1 + S_2$$

where

(4.27) $\qquad S_1 = \dfrac{3\pi}{4\lambda}(\partial_1^3 \hat{X}_0^1(0,0) + \partial_1 \partial_2^2 \hat{X}_0^1(0,0) + \partial_1^2 \partial_2 \hat{X}_0^2(0,0) + \partial_2^3 \hat{X}_0^2(0,0)),$

$$
\begin{aligned}
S_2 = \frac{3\pi}{4\lambda^2}(&-\partial_1^2 \hat{X}_0^1(0,0) \cdot \partial_1 \partial_2 \hat{X}_0^1(0,0) \\
&+ \partial_2^2 \hat{X}_0^2(0,0) \cdot \partial_1 \partial_2 \hat{X}_0^2(0,0) \\
&+ \partial_1^2 \hat{X}_0^2(0,0) \cdot \partial_1 \partial_2 \hat{X}_0^2(0,0) \\
&- \partial_2^2 \hat{X}_0^1(0,0) \cdot \partial_1 \partial_2 \hat{X}_0^1(0,0) \\
&+ \partial_1^2 \hat{X}_0^1(0,0) \cdot \partial_1^2 \hat{X}_0^2(0,0) \\
&- \partial_2^2 \hat{X}_0^1(0,0) \cdot \partial_2^2 \hat{X}_0^2(0,0)).
\end{aligned}
$$

(4.28)

Here $\hat{X}_0$ is the projection of the vector field X_0 restricted to the center manifold at 0 on the spectral subspace of $dX_0(0)$ corresponding to $\{i\lambda, -i\lambda\}$ and parallel to the spectral subspace corresponding to $\{-a_{22}\}$. This means that in the w coordinates

(4.29) $\qquad \hat{X}_0^i(w_1, w_2) = X_0^i(w_1, w_2, f(w_1, w_2)), \qquad i = 1, 2,$

where $\{w_1, w_2, f(w_1, w_2)\}$ is the center manifold at 0 (see [MMcC, section 4]). The derivatives of $\hat{X}_0$ are to be computed from those of X_0 according to the relations

(4.30) $\qquad \partial_k \partial_j \hat{X}_0^i(0,0) = \partial_k \partial_j X_0^i(0), \qquad \text{for} \quad i, j, k = 1, 2$

$$
\begin{aligned}
\partial_l \partial_k \partial_j \hat{X}_0^i(0,0) = \ &\partial_l \partial_k \partial_j X_0^i(0) + \partial_3 \partial_j X_0^i(0) \cdot \partial_l \partial_k f(0,0) \\
&+ \partial_3 \partial_k X_0^i(0) \cdot \partial_l \partial_j f(0,0) \\
&+ \partial_3 \partial_l X_0^i(0) \cdot \partial_k \partial_j f(0,0), \text{ for } i, j, k, l = 1, 2.
\end{aligned}
$$

(4.31)

In the expressions (4.31), $\partial_i \partial_j f(0,0)$ are to be filled in as they are given by
(4.32)
$$
\begin{pmatrix} \partial_1^2 f(0,0) \\ \partial_1 \partial_2 f(0,0) \\ \partial_2^2 f(0,0) \end{pmatrix} = -\Delta^{-1} \cdot \begin{pmatrix} 2\lambda^2 + a_{22}^2 & 2\lambda a_{22} & 2\lambda^2 \\ -\lambda a_{22} & a_{22}^2 & \lambda a_{22} \\ 2\lambda^2 & -2\lambda a_{22} & 2\lambda^2 + a_{22}^2 \end{pmatrix} \begin{pmatrix} \partial_1^2 X_0^3(0) \\ \partial_1 \partial_2 X_0^3(0) \\ \partial_2^2 X_0^3(0) \end{pmatrix}
$$

with

(4.33) $\qquad \Delta = -a_{22}(a_{22}^2 + 4\lambda^2).$

Remark that in our case $\partial_l \partial_k \partial_j X_0^i(0) = 0 \qquad \forall i, j, k, l.$

We preferred to get first the final expression of S_1 and S_2 respectively in the matrix components r_{ij}, using (4.25), and next in that expression to fill in the formulae of r_{ij}.

In this way S_1 depends only on $r_{13}, r_{23}, r_{31}, r_{32}$. As the definition (4.23) yields
(4.34)
$$\begin{cases} r_{13} = -\frac{1}{a_{11}(\lambda^2+a_{22}^2)} \cdot a_{22}, \\ r_{23} = -\frac{1}{a_{11}(\lambda^2+a_{22}^2)} \cdot \lambda, \\ r_{31} = -\frac{1}{a_{11}(\lambda^2+a_{22}^2)} \{a_{22}x^2 + 2a_{12}\lambda^2 x + a_{11}(\lambda^2 + a_{22}^2)[(\xi^2+1)a_{11}a_{22} + \lambda^2]\} \\ r_{32} = -\frac{1}{a_{11}(\lambda^2+a_{22}^2)} \cdot \lambda\{2a_{12}a_{22}x + 4a_{12}^2\lambda^2 + a_{11}(\lambda^2 + a_{22}^2)[(\xi^2+1)a_{11} - a_{22}]\}, \end{cases}$$

we obtain
(4.35)
$$S_1 = -\frac{3\pi}{2} \cdot \frac{\xi^2}{\lambda a_{22}(a_{22}^2 + 4\lambda^2)(\lambda^2 + a_{22}^2)} \{3a_{22}^2(a_{22}^2 + 2\lambda^2)x^2$$
$$+ 4\lambda^2 a_{12}a_{22}(4a_{22}^2 + 7\lambda^2)x + a_{11}(\lambda^2 + a_{22}^2)^2[(\xi^2+1)a_{11}(3a_{22}^2 + 8\lambda^2) - 2a_{22}\lambda^2]$$
$$+ 4a_{12}^2\lambda^4(5a_{22}^2 + 8\lambda^2)\}.$$

Analogously, S_2 depends only on $r_{11}, r_{12}, r_{21}, r_{22}$ whose expressions are:

(4.36)
$$\begin{cases} r_{11} = -\frac{1}{a_{11}(\lambda^2+a_{22}^2)} \cdot a_{22}x, \\ r_{12} = -\frac{1}{a_{11}(\lambda^2+a_{22}^2)} \cdot 2\lambda a_{12}a_{22}, \\ r_{21} = -\frac{1}{a_{11}(\lambda^2+a_{22}^2)} \cdot \lambda x, \\ r_{22} = -\frac{1}{a_{11}(\lambda^2+a_{22}^2)} \cdot 2\lambda^2 a_{12} \end{cases}$$

Thus

(4.37)
$$S_2 = \frac{3\pi}{2} \cdot \frac{\xi^2}{\lambda(\lambda^2 + a_{22}^2)}[a_{22}x^2 + 2a_{12}(\lambda^2 - a_{22}^2)x - 4\lambda^2 a_{12}^2 a_{22}].$$

The final expression of the stability coefficient is
(4.38)
$$V'''(0) \equiv S_1 + S_2 = -\frac{3\pi}{2} \cdot \frac{\xi^2}{\lambda a_{22}(a_{22}^2 + 4\lambda^2)} \{2a_{22}^2 x^2 + 2a_{12}a_{22}(10\lambda^2 + a_{22}^2)x$$
$$+ a_{11}(\lambda^2 + a_{22}^2)[(\xi^2+1)a_{11}(3a_{22}^2 + 8\lambda^2) - 2a_{22}\lambda_2]$$
$$+ 4a_{12}^2\lambda^2(8\lambda^2 + a_{22}^2)\}.$$

The formula (4.20), where $a_{11}\xi > 0$, shows that x can be taken as affine coordinate on the straight line δ. The expression of $V'''(0)$ shows then that far away on δ the stability coefficient is always negative, the Hopf bifurcation being thus supercritical and stable. In the case that the discriminant of the polynomial in x from the right hand side of (4.38) is positive – and this is precisely the condition (4.12) – there are points on δ where $V'''(0) > 0$; around such a point the bifurcation is subcritical and unstable.

The theorem is proved.

$\square$

Remark 4.4. As the left had side of the inequality (4.12) does not depend on a_{12}, it can be fulfilled by taking $|a_{12}|$ great enough.

On the other hand, for $\frac{a_{33}}{a_{22}}$ and $\frac{a_{22}}{a_{11}}$ great enough, the left hand side of (4.12) becomes negative, the inequality being thus satisfied independently of the value of a_{12} □

It would have been nice to show that the bifurcation is always supercritical and stable when the crossing point with $\sum_+$ lies outside a bounded set of $\sum_+$. But to lead and to interpret the calculations in a general point of $\sum_+$ new ideas are needed.

REFERENCES

[A] V. I. ARNOLD, *Les méthods mathématiques de la mécanique classique*, Ed. Mir, Moscow, 1976.

[CF] P. CONSTANTIN, C. FOIAS, *Navier-Stokes Equations*, University of Chicago Press, 1988.

[GG] M. GOTO, F. D. GROSSHANS, *Semisimple Lie Algebras*, M. Dekker, New York and Basel, 1978.

[MMcC] J. E. MARSDEN, M. McCRACKEN, *The Hopf Bifurcation and Its Applications*, Springer Verlag, New York, 1976.

[M1] G. MINEA, *Remarques sur P'unicité de la solution stationnaire d'une équation de type Navier-Stokes*, Rev. Roum. Math. Pures et Appl., tome XXI, no 8, 1976, pp. 1071-1075.

[M2] G. MINEA, *Sur les dérivées des opérateurs du type Navier-Stokes*, Annali di Mat. pura ed appl, CXXIX, 1981, pp. 131-142.

[T] R. TEMAM, *Infinite Dimensional Dynamical Systems in Mechanics and Physics*, Appl. Math. Sci., Springer Verlag, New York, 1988.

TURBULENT BURSTS, INERTIAL SETS
AND SYMMETRY-BREAKING HOMOCLINIC CYCLES
IN PERIODIC NAVIER-STOKES FLOWS

BASIL NICOLAENKO* AND ZHEN-SU SHE†

Abstract. We investigate bursting regimes of two-dimensional Kolmogorov flows. We link these dynamics with symmetry-breaking heteroclinic connections which generate persistent homoclinic cycles. Small-scale turbulent dynamics prevail in a neighborhood of these heteroclinic connections, while large-scale dynamics are associated to hyperbolic tori. These intermittent turbulent regimes are a prime example of dynamics on an inertial set (or exponential attractor) of the Navier-Stokes equations.

1. Introduction. Chaotic and turbulent flows can be considered as a dynamical system with very large numbers of degrees of freedom. The long time dynamics of these systems is related to the geometry of their global attractors. The global attractor is a compact, invariant, fractal set in the phase space to which all trajectories ultimately converge. 2D Navier-Stokes equations are indeed vested with a global attractor with finite fractal dimension [1]. Two questions important both physically and numerically are still outstanding: first is it possible to describe chaotic dynamics for the infinite dimensional Navier-Stokes system using a finite dimensional approximation (such as the usual Galerkin and/or pseudo-spectral schemes). To that effect, considerable endeavor has focussed on constructing *inertial manifolds*; these are finite-dimensional Lipschitz manifolds (in fact graphs) that attract all solutions at a uniform exponential rate, and such manifolds are invariant under the forward dynamics (positive semi-flow) [2,3]. For $2D$ Navier-Stokes equations describing the evolution of turbulent flows, the existence of an inertial *manifold* is still open.

The second relevant problem is how fast does the global attractor attract Navier-Stokes turbulent trajectories? The *rate of attraction need not be exponential*; in fact, physicists often voice objections at the mathematical global attractor; specifically, observed turbulent trajectories could be only slow transients, and only a vicinity of the global attractor be visited on experimental time scales. Paradoxically, as phrased by S. Smale, we cannot prove mathematically that 2D turbulence does not decay ultimately to a laminar regime (steady, periodic, quasi-periodic states). In that case, the global attractor would be relatively simple and yet we are still basically interested in the turbulent dynamics, be they transients.

Recently, a new set was introduced (see [4] and [5]) in order to overcome some of the theoretical difficulties that are associated with these issues; this new set is called inertial set. The inertial set is still a finite dimensional object imbedded in the infinite dimensional phase space. An inertial set, by definition, contains the

*Department of Mathematics, Arizona State University, Tempe, AZ 85287 and Center for Nonlinear Studies, Los Alamos National Laboratory, Los Alamos, NM 87545

†Applied and Computational Mathematics, Princeton University, Princeton, NJ 08544

global attractor and attracts all trajectories at a uniform exponential rate. Consequently, it contains the slow transients as well as the global attractor. In the theory of dynamical systems the slow transients correspond to slowly converging stable manifolds that are in some sense close to central manifolds. Numerical simulations of infinite-dimensional dynamical systems often capture both the slow transients and parts of the attractor. After a large but finite time the state of the system obtained from the numerical calculation may often be a finite distance from the global attractor but of infinitesimal distance to the inertial set. In this sense, we call the inertial set an Exponential Attractor [20] to be consistent with the physical intuition. More specifically, after a short transient time the infinite-dimensional system is arbitrarily close to an exponential attractor; the existence of such an inertial set has been demonstrated for 2D Navier-Stokes flows [4,5], through geometric properties of the trajectories.

We present here a study of a specific Navier-Stokes flow system, the two-dimensional Kolmogorov flow, for which we conjecture that transition from weak chaos to moderate turbulence takes place on a relatively low-dimensional component of such an exponential attractor. This is a flow where an enhanced transport phenomenon occurs intermittently related to symmetry-breaking heteroclinic orbits. This is a moderate, but not small Reynolds number system at the onset of a substantially developed turbulent regime. It turns out that a much more random motion is developed during enhancement of transport. The states on which the system pass most of the time correspond in fact to much more spatially organized patterns. We will show that heteroclinic orbits connect those equivariant states under symmetry groups of the equation. Globally these heteroclinic orbits generate persistent homoclinic cycles.

This is a prime situation were enhanced turbulence takes place on *an inertial set, rather than on the attractor*. The attractor contains the geometric homoclinic cycle. The bursting dynamics takes place in a *tubular neighborhood of such a cycle* and are slowly attracted to it. Dynamics on the *ideal* cycle itself would take an infinite time. Yet "subgrid" *noise* drastically slows down the rate of attraction, and in practice, the turbulent dynamics persist on a very long time-scale in a small tubular neighborhood of the limit cycle; this neighborhood does lie on an inertial set of finite fractal dimension and contains a slowly convergent piece of the otherwise infinite dimensional stable manifold of the hyperbolic states.

2. The Kolmogorov flow: Bursting regimes. The two-dimensional Kolmogorov flow is the solution of the 2-D Navier-Stokes equation with a uni-directional force $\mathbf{f} = (\nu k_f^3 \sin k_f y, 0)$. It was introduced by Kolmogorov in the late fifties as an example on which to study transition to turbulence. For large enough viscosity ν, the only stable flow is a plane parallel periodic shear flow $\mathbf{u}_0 = (\nu k_f \sin k_f y, 0)$, usually called the "basic Kolmogorov flow". The macroscopic Reynolds number of the basic flow is easily found to be $1/\nu$; this will be used later as a free parameter to define the bifurcation sequence. It was shown by Meshalkin and Sinai [6] that large-scale instabilities are present for Reynolds numbers exceeding a critical value, $\sqrt{2}$. This large-scale instability has been shown by Nepomnyachtchyi [7] and Sivashinsky [8] to be of negative-viscosity type, in the sense that the basic anisotropic flow generates a negative viscosity for large scale perturbations.

In a 2π-periodic box, the equations are:

(1)
$$\frac{\partial \mathbf{u}}{\partial t} + \mathbf{u} \cdot \nabla \mathbf{u} + \nabla p = \nu \nabla^2 \mathbf{u} + \mathbf{f}, \quad \nabla \cdot \mathbf{u} = 0,$$
$$\mathbf{f} = (\nu k_f^3 \sin k_f y, 0), 0 \leq x, y \leq 2\pi.$$

Sequences of bifurcations have been investigated by She [9,10]. But the most interesting transitions occur at even higher Reynolds number and are the subject of this study; they lead to sparsely distributed bursts in time for a fairly large range of Reynolds number above a certain threshold, i.e. $Re \approx 20.8$ for $k = 8$. The most striking feature of this transition is that the bursts generate substantial spatial disorder and drive developed turbulence.

In Figure 1, we show the time evolution of the maximal (pointwise) vorticity $\omega = \text{curl}\,\mathbf{u}$ of the 2-D Navier-Stokes equations for a Reynolds number equal to 20.8, nearly at the threshold of the transition to the intermittent scenario. Here the mode $(k_x, k_y) = (0, 8)$ is the forcing mode corresponding to a small scale y-dependent shear flow. It can be seen that the bursting event is characterized by an energy depletion of the large scale modes, $(1, 0)$ and $(0, 1)$ and feeding of the small scale ones [11].

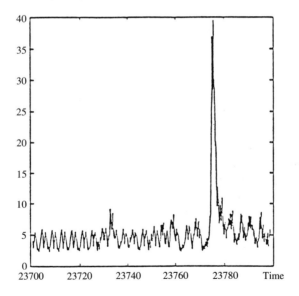

Fig. 1: A strong homoclinic excursion at Re=20.8, time = 23775. Vertical axis represents maximum vorticity.

Figure 1 highlights the typical temporal dynamics of the vorticity. The dynamics follow a long laminar regime, then undergo a strong turbulent (chaotic) explosion; then seem to settle down to a laminar regime at the same level as before; then another explosion follows. Intervals between explosions are not *constant* and fluctuate randomly. Careful study of Fourier modes [11] evidence that:

- the dynamics of the laminar regimes are on metastable T_2 Tori (laminar windows).

- successive "plateaus" *do not correspond to identical dynamical states*, but rather to sequences of states $S_1, S_2, S_3 \ldots$ mapped under some subgroup action Λ. This will be demonstrated by a phase analysis of the complex Fourier coefficients, but can already be evidenced by comparing the vorticity fields before (Figure 4) and after (Figure 5) vorticity explosion.

- spatial order is destroyed during the burst (Figure 6), with spatial dynamics on a *much smaller space-scale*; spatial chaos coexists with temporal chaos, which is a good characterization of turbulence.

Clearly the vortices after the burst (Figure 5) have been shifted w.r.t. the vortices before the burst (Figure 4), both in the x- and y-directions. Angular phase analysis of complex Fourier modes (Figures 2 and 3) demonstrate definite phase shocks at the burst, as opposed to slow-scale phase drift (mode $k_x = 1$, $k_y = 0$) or steady levels (mode $k_x = 0$, $k_y = 1$) during the laminar "plateaus." One of the frequencies of the hyperbolic T_2 state corresponds, indeed, to constant drift along the x-axis. Careful analysis of the angular phase shocks pins down the exact subgroups Λ under which vortex states during laminar "plateaus" between bursts are mapped. Recall that if $\psi_k(t)$ is the complex Fourier coefficient of a mode $k = (k_x, k_y)$, the phase is defined as:

$$\psi_k(t) = |\psi| e^{i\varphi_k},$$
$$\varphi_k(t) = \arctan(\mathrm{Im}\psi_k(t)/\mathrm{Re}\psi_k(t))$$

In Fig. 2, the phase jump of the mode $(1,0)$ is $+\pi/2$, whereas (Fig. 3) the phase jump of $(0,1)$ is $-7\pi/8$.

In what follows we will describe this behavior in terms of heteroclinic connections and homoclinic loops in phase space and discuss the connection with invariant symmetry groups of the system. The heteroclinic connections are structurally stable: the bursting regimes are evidenced up to and beyond $\mathrm{Re} = 100$ (for $k_f = 8$); intervals between bursts become shorter as Re $\uparrow$, so that the coherent vortices ultimately appear only during brief intermittencies, whereas the strong bursts dominate nearly all of the dynamics. It should be pointed out that the Reynolds number $\mathrm{Re} = 1/\nu$ is only a macroscopic number. A much better representative of small scale dynamics is the Grashoff Number:

$$(2) \qquad\qquad G_a = \frac{|f|^2}{\nu^2 \lambda_1},$$

where $|f|^2 = \int \int_{(0,2\pi)^2} |f|^2 dx\, dy$, and $\lambda_1 = 1$ is the first eigenvalue of the Stokes operator. It is known rigorously that the fractal dimension d_F of the global attractor $\mathcal{A}$ for 2D-Navier Stokes is bounded by [12]:

$$(3) \qquad\qquad d_F(\mathcal{A}) \le C G_a^{2/3}(1 + \log G_a)^{1/3}.$$

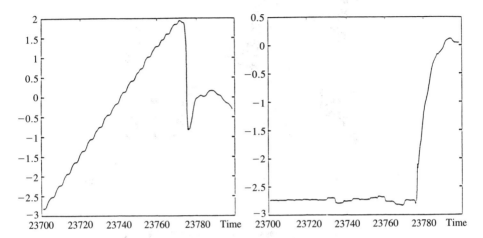

Fig. 2: Cumulative phase history for the phase of the complex Fourier mode $k_x = 1, k_y = 0$ (for burst in Fig. 1)

Fig. 3: Cumulative phase history for the phase of the complex Fourier mode $k_x = 0, k_y = 1$ (for burst in Fig. 1)

For our Kolmogorov flow with Re $\approx$ 50, $k_f = 8$ and the force given in (1), this yields $G_a \simeq 26,000$ and $d_F(\mathcal{A}) \leq 1900$. In fact, it has been shown that a better effective Reynolds number fully taking into account the small scale $2\pi/k_f$ is:

$$(4) \qquad \qquad \mathrm{Re}_m \sim (G_a)^{2/3}$$

which gives $\mathrm{Re}_m \sim 860$ for $1/\nu = 50$. Our system has some one thousand degrees of freedom on the attractor, which are indeed explored by smaller scale dynamics during the strong vorticity bursts. During the intermittent "plateaus" where dynamics stay topologically very close to a T_2 torus, the dimensionality of the dynamics is considerably lower, with energy concentrated in the largest scale modes. The homoclinic cycle allows for a dynamic mechanism of intermittencies between large-scale, low dimensional dynamics, and small-scale, large-dimensional ones (with $\mathcal{O}(1000)$ degrees of freedom).

3. Dynamics of the symmetry breaking homoclinic cycles. The intermittent chaotic (turbulent) behavior of the Kolmogorov flow is deeply connected with its groups of symmetries and related symmetry breakings. Specifically, the Kolmogorov flow is equivariant under the group

$$(5) \qquad \qquad \Gamma = O(2) \times D_k,$$

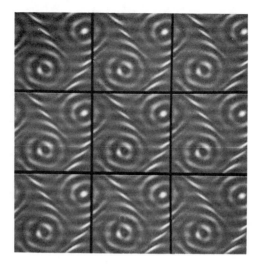

Figure 4: The laminar system of vortices before the burst in Fig. 1.

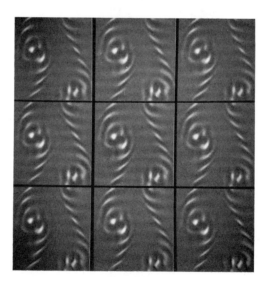

Figure 5: The laminar system of vortices after the burst in Fig. 1. Notice the shifts.

where the first group acts on x, and the second on y, with $k \equiv k_f$. Precisely:

(i) $O(2)$ equivariance in x is arbitrarily shift invariance, and invariance under reflection $x \to -x$.

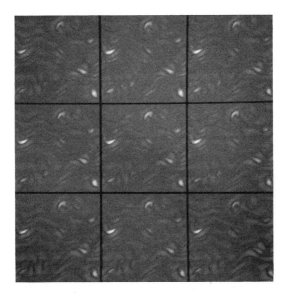

Figure 6: Both spatial and temporal disorder at small scales prevail during a burst.

(ii) discrete D_k equivariance in y is invariance under shifts $\pm 2\pi/k_f$ in y, as well as invariance under reflection $y \to -y$.

In terms of irreducible group actions on the stream function $\Psi(x, y)$, this can be characterized by

(i) $O(2) \times Z(2)$ equivariance in (x, y):

$$x \to -(x + \theta), \text{ for every } \theta,$$
$$(6) \qquad y \to -y,$$
$$\psi(x, y) \to +\psi(x + \theta, y);$$

(ii)

$$x \to x, y \to y + 2n\pi/k_f,$$

$$(7) \qquad \psi \to \psi$$

(iii) Symmetry axes in y:

$$y \to \pi/k_f - y + 2n\pi/k_f, \quad \text{for all} \quad n,$$
$$(8) \qquad x \to x,$$
$$\psi(x, y) \to -\psi(x, y);$$

for the above, the equation for the stream function is written as:

$$(9) \qquad \frac{\partial}{\partial t}\Delta\psi + \frac{\partial\psi}{\partial y}\frac{\partial}{\partial x}(\Delta x) - \frac{\partial\psi}{\partial y}\frac{\partial}{\partial y}(\Delta\psi) = \frac{1}{\Re}\Delta^2\psi + \nu k_f^4\cos(k_f y)$$

(that is, the force for the Navier-Stokes equation is chosen as $(\nu k_f^3\sin k_f y, 0)$).

The metastable large scale vortex systems during the laminar "plateaus" (Figure 1) are associated to hyperbolic T_2 tori which which can be shown to be equivariant under a smaller subgroup Λ_0 of Γ. Let $S_1, S_2, \ldots, S_k$ be the flow invariant Tori observed between the bursts. A *heteroclinic cycle* of the S_j's is a collection of paths which connect each S_j to S_{j+1}, starting on the unstable manifold of S_j and ending on the stable manifold of S_{j+1}; here S_{k+1} is identified with S_1. For symmetric systems, heteroclinic cycles between saddle points can be structurally stable. In the context of $O(2)$ symmetries, this was established in [13] for Kuramoto-Sivashinsky equations and [14] for a more general $O(2)$ context. Several extensions have been worked out for Toris in $O(2)$ and Steady States and Torus in $O(3)$ [15], as well as a global D_4 symmetry [16]. The complexity of the Kolmogorov flow stems from its $O(2) \times D_k$ symmetry, k arbitrary.

Symbolize the evolution problem (1) in an abstract space X as:

$$(10) \qquad \frac{\partial\psi}{\partial t} = N(\psi),$$

where $\psi(x, y, t)$ is the stream function associated to the velocity field $\mathbf{u}(x, y, t)$ in (1).

To understand the symmetry breaking mechanisms involved, we review some basic facts of group equivariance. The Kolmogorov flow is equivariant under Γ (as defined in 5) if:

$$(11) \qquad N(\gamma\psi) = \gamma N(\psi),$$

for every $\gamma \in \Gamma$, $\psi \in X$. This impacts on steady states, since for every x_0 such that $N(x_0) = 0$, the group orbit through x_0, defined as:

$$(12) \qquad \Gamma\psi_0 = \{\gamma\psi_0 : \gamma \in \Gamma\}$$

is also an orbit of steady states. We consider solutions in the same Γ-orbit as equivalent. Similar properties carry over to periodic solutions, provided that we consider the group $\Gamma \times S^1$ (normalizing the circle group S^1 to period 2π).

One defines the symmetries of any particular hyperbolic state x_0 through the *isotropy subgroup* of x_0 defined as:

$$(13) \qquad \Sigma(x_0) = \{\gamma \in \Gamma : \gamma x_0 = x_0\},$$

and $\Sigma(x_0 :) \subset \Gamma(\text{resp.}\Sigma(x_0) \times S^1$ for periodic solutions $x_0(t))$. In the context of the homoclinic cycles investigated here, typically $\Sigma(x_0)$ for the homoclinic critical points x_0 are $D_p \times D_q$, for some $p \geq 2, q \geq k_f$; that is, the diedral groups of rotations of $2\pi/p$ in x, $2\pi/q$ in y, together with reflections in x and y.

Given such an isotropy subgroup $\Sigma(\mathrm{resp.}\Sigma \times S^1)$, an important linear subspace of X is the fixed point subspace of Σ:

$$(14) \qquad \mathrm{Fix} \ (\Sigma) = \{y \in X : \sigma y = y \text{ for all } \sigma \in \Sigma\}.$$

The importance of fixed point subspaces lies in that Γ-equivariance of $N(x)$ leaves Fix (Σ) *invariant*; $\sigma N(x) = N(\sigma x) = N(x)$ for every $\sigma \in \Sigma \subset \Gamma$, so that $N(x) \in \mathrm{Fix}(\Sigma)$. This enables one to restrict the investigation of symmetry breaking bifurcations (w.r.t. Σ) to fixed point subspaces.

Figures 4 and 5 reveal the group of symmetries $\Lambda_0 \subset D_2 \times D_2$ of the metastable Tori T_2 between the bursts. Recall that D_2 action on the x-axis (respectively, y-axis) is represented by $x \to -x$ (respectively $y \to -y$) and $x \to x + \pi$ (respectively $y \to y + \pi$). In a *moving* coordinate system (constant phase shift along the x-axis, as the T_2 torus travels with constant speed parallel to the x-direction) and centering the origin at the eye (center) of any of the spatial eddies structures, we have invariance under the 4 elements discrete group Λ_0:

$$\gamma_i : x \to x, y \to y \quad \text{(identity)}$$
$$\gamma_i : x \to -x, y \to -y \quad \text{(center of symmetry)}$$
$$\gamma_2 : x \to \pi - x, y \to \pi + y$$
$$\text{(translation by the vector } \pi\mathbf{e}_x + \pi\mathbf{e}_y, \text{followed by symmetry}$$
$$\text{with respect to the axis } x = \pi)$$
$$\gamma_3 : x \to \pi + x, y \to \pi - y$$
$$\text{(translation by the vector } \pi\mathbf{e}_x + \pi\mathbf{e}_y, \text{followed by symmetry}$$
$$\text{with respect to the axis } y = \pi$$

The group Λ_0 is in fact isomorphic to the group D_4 of symmetries of the square, through a $45°$ rotation of the coordinate axis x and y, as clearly seen from figures 4 and 5.

From a careful study of the symmetries of the hyperbolic tori between bursts in the Kolmogorov flow, we determine that, for $k_f = 8$:

(i) Each hyperbolic torus S_i admits the same isotropy subgroup $\Lambda_0 \subset D_2 \times D_2$, where Λ_0 is a *maximal* subgroup.

(ii) Fix (Λ_0) defines a flow invariant line, which contains all the S_i.

(iii) Each S_i is mapped into S_{i+1} by action of a maximal subgroup Λ, with $\Lambda_0 \subset \Lambda \subset \mathcal{O}(2) \times D_8$. More generally, any S_j is mapped into any S_k by action of an element of

$$(15) \qquad \Lambda = D_4 \times D_{16}$$

(group of symmetries of the square along x, times the group of the 16-sided regular polygone along y).

Property (iii) is fundamental. It must be verified computationally over very long time series for Re very close to bifurcation (see Fig. 2 and 3). The phase "shocks" determine the actual elements of $D_4 \times D_{16}$ which map an S_j into the next S_{j+1}. In the x-direction, we have observed shifts of $\pm\frac{\pi}{2}, \pm\pi$ and $\frac{3\pi}{2}$. In the y-direction, we have observed shifts of $\pm\frac{\pi}{8}, \pm\frac{2\pi}{8}, \pm\frac{3\pi}{8}, \pm\frac{4\pi}{8}, \pm\frac{5\pi}{8}, \pm\frac{6\pi}{8}, \pm\frac{7\pi}{8}$ and $\frac{12\pi}{8}$ (higher order shifts are actually present, because of the symmetry $\varphi \to 2\pi - \varphi$). From the odd multiples of $\frac{\pi}{8}$ in the above list, we conclude that we are *actually dealing with* D_{16}, and not just $D_2 \times D_8$. Moreover:

(iv) both the *unstable linear eigenspace* $E^u(S_i)$ at S_i, the *unstable manifold* $W^u(S_i)$ at S_i and Fix (Λ_0) are contained in an invariant linear space $\mathcal{A}_{i,j}$ of the Kolmogorov flow:

(16a) $$\text{Fix}\,(\Lambda_0) \oplus E^u(S_i) \subset \mathcal{A}_{ij},$$

(16b) $$W^u(S_i) \not\subset \mathcal{A}_{ij}$$

(v) $\mathcal{A}_{ij}$ is the fixed point subspace of a submaximal isotropy subgroup T:

(17) $$T \subset \Lambda_0 \subset \Lambda \subset \Gamma.$$

Let us clarify properties of $E^u(S_i)$ and $W^u(S_i)$. Let S_1 be a basic hyperbolic state (generalized saddle) corresponding to a plateau. Its isotropy subgroup is $\Lambda_0 \subset \Gamma$. Writing the Navier-Stokes equations as $\frac{\partial\psi}{\partial t} = N(\psi)$, the linearized operator $dN(S_1)$ is equivariant under Γ; for every $\gamma \in \Gamma$, every $x \in X$:

(18a) $$dN(\gamma S_1)\gamma_x = \gamma dN(S_1)x,$$

and dN also commutes with Λ_0, the isotropy subgroup:

(18b) $$dN(\sigma S_1)\sigma x = \sigma dN(S_1)x,$$

for every $\sigma \in \Lambda_0, x \in X$. Classically, [17] implies that *we can decompose the space* as:

(19) $$X = W_1 \oplus W_2 \oplus \cdots \oplus W_k,$$

where the W_1 *are irreducible* and

(20) $$dN(S_1)(W_j) \subset W_j;$$

moreover, we can always take

(21a) $$W_1 = \text{Fix}(\Lambda_0)$$

Symmetry breaking implies that the *unstable* linear eigenspace at $S_1, E^u(S_1)$ is included in some $W_j \neq \text{Fix}\,(\Lambda_0)$, say:

(21b) $$E^u(S_1) \subset W_2.$$

Now

(22) $$\text{Fix}(\Lambda_0) \oplus W_2 = A_{1,2}$$

generates *an invariant (linear) manifold $A_{1,2}$ for the Navier-Stokes flow*, that is, if $x(t_0) \in A_{1,2}$ then $x(t) \in A_{1,2}$ for all $t \geq t_0$. Of course:

(23) $$\text{Fix}(\Lambda_0) \oplus E^u(S_1) \subset A_{1,2}.$$

Classically, any invariant manifold A is equivariant under Γ, that is, γA is also an invariant manifold. It follows that the (nonlinear) unstable manifold $W^u(S_1)$ is contained in $A_{1,2}$:

(24) $$W^u(S_1) \subset A_{1,2}.$$

From the above, we establish the:

PROPOSITION. *In the invariant linear subspace $\mathcal{A}_{ij}$, the hyperbolic state S_i is a saddle, whereas the state S_{i+1} is a sink. In particular, $E^u(S_i)$ and $W^u(S_i)$ are mapped into $E^u(S_{i+1})$ and $W^u(S_{i+1})$ by action of the maximal subgroup $\Lambda = D_4 \times D_{16}$.*

To rigorously establish the chains of heteroclinic connections:

(25) $$S_1 \xrightarrow{\Lambda} S_2 \xrightarrow{\Lambda} \cdots S_j \xrightarrow{\Lambda} S_{j+1} \cdots$$

we must construct the lattice of isotropy subgroups of $O(2) \times D_8$ and establish permissible connections in Fix (Λ_0) through Fix (T), with such connections equivariant under $\Lambda = D_4 \times D_{16}$. Effective dimensionality of the fixed sets A_{ij} can be reduced with the methods developed in [15] for heteroclinic cycles involving periodic solutions with $O(2)$ symmetry. This work will be reported elsewhere.

4. Conclusion. In summary, we have shown that the intermittent bursting events in the two-dimensional Kolmogorov flow, a specific Navier-Stokes flow system, are connected to heteroclinic excursions from one spatially organized state to another, both are equivariant by symmetry groups $D_r \times D_s$. During the excursion, the system visits regions in phase space characterized by spatially chaotic structures which greatly enhance the transfer from large to small scales and the dissipation. There is no indication that heteroclinic excursion is confined in a low dimensional manifold; rather states of higher dimensions are involved which correspond to disordered spatial patterns during the burst. It is not clear yet how enhanced transport can be usually associated with coherent events as suggested by Newell et al. [19]. Boundary layer turbulence is an example of a flow system where strong shear motions dominate the dynamics. One of the important problems there is the turbulence production mechanism. It is believed that bursts near the wall inject energy into the outer regions, and thus are the main source of the turbulence production. It has been speculated that homoclinic excursions may be the dynamical origin of the bursting events in boundary layer flows. Indeed, Aubry et al. [18] have found that reduced dynamical systems obtained by projecting the boundary layer turbulence correlation function do contain homoclinic cycles. We have seen that the Kolmogorov flow displays, at high enough Reynolds number, temporally intermittent behavior. In this sense, the Kolmogorov flow provides an explicit example of a Navier-Stokes flow system exhibiting temporal intermittency, without any modelling. Moreover, intermittent bursting events in the Kolmogorov flow are intimately related to the generation of spatially disordered patterns, and thus are directly responsible for the production of developed turbulence.

These intermittent bursting dynamics are a prime example of dynamics on an exponentially attracting inertial set of the Navier Stokes equations. The attractor itself contains only the hyperbolic Tori S_i and the homoclinic cycles obtained from "chaining" (concatenating) the unstable manifolds:

$$(26) \qquad \qquad \text{cycle} = \bigcup_{(i)} W^u(S_i),$$

for any subset of all possible S_i. Yet the bursting dynamics take place within a tubular neighborhood of these cycles, with *very slow convergence* to the limit ideal cycle. In fact, in our computational simulation, we do observe only a marginal increase of the pseudo-period between the bursts. This is definitely not a theoretical deficiency of the pseudo-spectral $(512)^2$ Galerkin scheme. In fact, in [4,5], we establish the:

STABILITY THEOREM FOR THE GALERKIN APPROXIMATION SCHEME. *The Inertial Set for the Galerkin dynamical system with order $N(\epsilon)$ large enough is ϵ-close to the exact Navier-Stokes Inertial Set, modulo a time shift, where ϵ is a function of the scheme order, and vice-versa. Moreover, the Galerkin Inertial Set is ϵ-close to the exact Navier-Stokes global attractor (no time shift) and the exact Navier-Stokes Inertial Set if ϵ-close to the Galerkin global attractor.*

Such stability theorems are *not generally true for approximations of the global attractor* (for which only upper semi-continuity holds). What does considerably slow down the convergence to the attractor in fact subgrid *noise* and slaved modes. Even for a trajectory very close to a hyperbolic Torus Si, noise can jitter the trajectory both into the unstable manifold $W^u(S_i)$ and into the stable manifold $W^s(S_i)$ further away from S_i. The very strong exponential attraction rate of the inertial set immediately relaxes the dynamics onto the inertial set after a brief transient. In fact, the larger the inertial set is, the faster the exponential attraction.

Indeed, exact geometric heteroclinic connections (concatenated into homoclinic cycles) are imposed by symmetry subgroups, which also split the unstable and stable manifolds into invariant components. These ideal connections define the (physical) attractor in an appropriate subspace, with a global tube of attraction for forward times in the global phase space. However, numerical observations beyond, yet very close to the bifurcation, show very slow convergence to these ideal heteroclinic connections. What does dominate is a very noisy connection (homoclinic chaos?) between the different metastable Tori. We can account for it by a change of the geometric structure of the stable-unstable manifolds. More precisely, some phase-space neighborhoods of the heteroclinic connections (far from the hyperbolic Tori) develop a more complex fractal structure and dynamically imply a much broader range of formerly slaved subgrid modes. The dynamics are then properly described by a much larger (topological) manifold. One might surmise that the geometric structure near the hyperbolic Tori changes very little, whereas the heteroclinic connection could be vested with a complex fractal structure, within its linear symmetry subspace. To such an extent formerly slaved subgrid modes initiate a dynamical rebellion.

Moreover, even if the connections have a complex fractal structure, the characteristic times of the heteroclinic excursions are determined by the local geometric properties of the hyperbolic Tori (ratios of the unstable eigenvalues to the smallest stable eigenvalues of the local linearization). The theoretical convergence to the ideal heteroclinic connections (global attractor) should be reflected into slower and slower bursts in an increasing geometric progression. Still we do not observe anything alike, even with increased numerical resolution, and the time intervals between bursts are randomly spaced around some mean. This is a strong indication that our system is not attracted exponentially by the global attractor, and we conjecture that the observed dynamics are in fact controlled by an *Exponential Attractor* which enlarges the global attractor.

Clearly, the dynamics involve a larger piece (*"slow stable manifold"*) of the ∞-dimensional $W^s(S_i)$, rather than the low dimensional $W^S(S_i) \cap W^u(S_{i-1})$. *One could conjecture that both experimentally and computationally observed 2D Navier-Stokes Turbulence actually takes place on Navier-Stokes Inertial Sets, as irreducible subgrid noise prevents full relaxation to the ideal global attractor.*

Acknowledgements. This research is supported by the Air Force Office for Scientific Research through its URI Grant to the Mathematics Department at Arizona State University and by DARPA under contract N00014-86-K-0769. We also wish to acknowledge the support and the hospitality of the Center for Nonlinear Studies at Los Alamos National Laboratory. We are indebted to Randy Heiland for the computer graphics, and to P. Chossat for helpful discussions.

REFERENCES

[1] R. TEMAM, *Infinite Dimensional Dynamical Systems in Mechanics and Physics*, Applied Mathematics Series, Vol. 68 (Springer, Berlin, 1988).

[2] C. FOIAS, G. SELL AND R. TEMAM, *Inertial manifolds for nonlinear evolutionary equations*, 73 (1988), pp. 309–353.

[3] P. CONSTANTIN, C. FOIAS, B. NICOLAENKO AND R. TEMAM, *Integral Manifolds and Inertial Manifolds for Dissipative Partial Differential Equations*, Appl. Math. Sciences, no. 70 (Springer, New York, 1988).

[4] A. EDEN, C. FOIAS, B. NICOLAENKO AND R. TEMAM, *Inertial sets for dissipative evolution equations*, monograph in preparation.

[5] A. EDEN, C. FOIAS, B. NICOLAENKO AND R. TEMAM, *Inertial sets for dissipative evolution equations*, IMA Preprint no. 694 (1990).

[6]   MESHALKIN, L.D. & SINAI, YA. G., *J. Appl. Math.*, (PMM), 25, 1700 (1961).

[7]   NEPOMNYACHTCHYI, A.A., *Prikl. Math. Makh.*, 40(5), 886 (1976).

[8]   SIVASHINSKY, G.I., *Physica*, 17D, 243 (1985).

[9]   SHE, Z.-S, *Proc. on Current trends in turbulence research*, AIAA, 1988.

[10] SHE, Z.-S., *Phys. Lett. A*, 124, 161 (1987).

[11] NICOLAENKO, B. AND SHE, Z.-S., *Temporal Intermittency and Turbulence Production in the Kolmogorov Flow*, in *Topological Fluid Mechanics*, Cambridge Univ. Press, 1990, pp. 256–277.

[12] CONSTANTIN, P. AND FOIAS, C., *Navier Stokes Equations*, Univ. of Chicago Lectures in Mathematics, 1989, pp. 256–277.

[13] KEVREKIDIS, I., NICOLAENKO, B. AND SCOVEL, C., *Back in the Saddle Again: A Computer Assisted Study of the Kuramoto-Sivashinsky Equation*, SIAM J. Appl. Math., Vol. 90, 3 pages 760–790 (1990).

[14] ARMBRUSTER, D. GUCKENHEIMER, J. & HOLMES, PH.,, *Physica D.*, (1989).

[15] MELBOURNE, I., CHOSSAT, P. AND GOLUBITSKY, M., *Heteroclinic Cycles involving Periodic Solutions in Mode Interactions with $O(2)$ Symmetry*, to appear; Also, Armbruster, D. and Chossat, P., Heteroclinic cycles in Mode Interaction with $O(3)$ Symmetry, to appear.

[16] GUCKENHEIMER, J., *Square Symmetry in Binary Convection*, to appear.

[17] GOLUBITSKY, M. , STEWART, I., SCHAEFFER, D.G., *Singularities and Groups in Bifurcation Theory*, Volume II, Springer-Verlag Ed., 1988.

[18] AUBRY, N., HOLMES, P., LUMLEY, J.L. & STONE, E., J. Fluid Mech., 192, 112 (1987).

[19] NEWELL, A., RAND, D., *Physics Letters*, A (1988).

[20] EDEN, A., FOIAS, C. NICOLAENKO, B. AND SHE, Z.S., *Exponential Attractors and their Relevance to Fluid Dynamics Systems*, submitted to Physica D.

NAVIER-STOKES EQUATIONS IN THIN 3D DOMAINS III: EXISTENCE OF A GLOBAL ATTRACTOR

GENEVIÈVE RAUGEL AND GEORGE R. SELL

1. Introduction.

Over 50 years ago, Leray (1933, 1934a, 1934b) published his pioneering works on the Navier-Stokes equations which led to the modern mathematical theory of fluid dynamics. These equations describe the time evolution of solutions of mathematical models of viscous incompressible fluid flows. Since the solutions of these equations depend on both space and time, one is especially interested in the phenomenon of the time evolution of the spatial variations of the solutions. This phenomenon, which is described with more precision later, is referred to as the *regularity* of solutions and it is one of our concerns here. During the last few years, much attention has also been focused on the study of attractors for the Navier-Stokes equations. This is related to a new insight in turbulence, which relies two theories of turbulence, the conventional theory of turbulence and the dynamical system approach (see, for instance, Babin and Vishik (1983), Constantin and Foias (1985), Constantin, Foias and Temam (1985), Constantin, Foias, Manley and Temam (1985), Foias and Temam (1987)).

The Navier-Stokes equations on a bounded region $\Omega \subset \mathbf{R}^n$, $n = 2, 3$, are given by

$$(1.1) \qquad U_t - \nu \Delta U + (U \cdot \nabla)U + \nabla P = F, \quad \nabla \cdot U = 0,$$

with various boundary conditions, where ∇ is the gradient operator and Δ is the Laplacian. In this paper we treat the case where $\Omega = \Omega_\varepsilon$ is a thin 3-dimensional domain, i.e., $\Omega_\varepsilon = Q_2 \times (0, \varepsilon)$, where ε is a small positive parameter, and Q_2 is a smooth bounded region in $\mathbf{R}^2$, or $Q_2 = (0, \ell_1) \times (0, \ell_2)$, $\ell_i > 0$, $i = 1, 2$. The equation (1.1) can be supplemented with various boundary conditions. When $Q_2 = (0, \ell_1) \times (0, \ell_2)$, the equation (1.1) can be supplemented with the periodic and zero mean conditions:

$$(1.2) \qquad \begin{cases} U|_{y_j=0} = U|_{y_j=\ell_j}, \quad j = 1, 2; \quad U|_{y_3=0} = U|_{y_3=\varepsilon} \\ \text{and} \\ \int_\Omega U \, dy = 0. \end{cases}$$

(In this case, one assumes that $\int_\Omega F \, dy = 0$).

In all these cases let V be the space of the divergence free $U \in H^1(\Omega)^n$ satisfying the respective boundary conditions, let H be the closure of V in $L^2(\Omega)^n$ and let $\mathbf{P}_n$ denote the orthogonal projection of $L^2(\Omega)^n$ onto H. Then the Navier-Stokes equations (1.1) on Ω can be written in the abstract form

$$(1.3) \qquad U' + \nu AU + B(U, U) = \mathbf{P}_n F,$$

where $AU = -\mathbf{P}_n \Delta U$, and $B(U_1, U_2) = \mathbf{P}_n(U_1 \cdot \nabla)U_2$. For the sake of simplicity we assume here that the forcing function $F = F(t)$ satisfies

$$(1.4) \qquad F(\cdot) \in L^\infty([0, \infty), L^2(\Omega)^n).$$

If U_0 belongs to H, by a weak Leray solution on the time interval $[0, T]$, we mean a function $U(\cdot) \in L^2(0, T; V) \cap L^\infty(0, T; H) \cap C_w^0([0, T]; H)$ satisfying

$$(1.5) \qquad \langle U', w \rangle + \langle \nu AU, w \rangle + \langle B(U, U), w \rangle = \langle \mathbf{P}_n F, w \rangle, \quad \text{a.e. in } t, \text{ for all } w \in V,$$

$$(1.6) \qquad U(0) = U_0,$$

and the energy inequality

$$(1.7) \qquad \frac{1}{2}\|U(t)\|_H^2 + \nu \int_{t_0}^t \|U(s)\|_V^2 \, ds \leq \frac{1}{2}\|U(t_0)\|_H^2 + \int_{t_0}^t \langle \mathbf{P}_n F, U(s) \rangle \, ds$$

for $t_0 = 0$ and for all $0 \leq t_0 \leq t \leq T, t_0, t$ a.e. in $[0, T]$. (We recall that $C_w^0([0, T]; H)$ is a subspace of $L^\infty(0, T; H)$ consisting of functions which are weakly continuous, that is, for each $h \in H$, the mapping $t \to \langle U(t), h \rangle$ is a continuous mapping. In particular, (1.6) is taken in this sense).

If U_0 belongs to V, by a *regular* or *strong* solution on the time interval $[0, T]$, we mean a function $U(\cdot) \in L^2(0, T; D(A)) \cap L^\infty(0, T; V) \cap C^0([0, T]; V)$ satisfying (1.5), and (1.6). Note that the strong solutions satisfy the energy inequality (1.7).

The study of the existence, uniqueness and regularity of solutions to the Navier-Stokes equations, both in 2-dimensions and 3-dimensions, has attracted a wide interest beginning with Leray (1933, 1934a, 1934b). We cannot give a history of this study here, but special mention should be made of the contributions of Hopf (1951), Lions and Prodi (1959), Serrin (1962), Fujita and Kato (1964), Ladyzhenskaya (1967, 1969), Masuda (1967), Komatsu (1980), and Caffarelli, Kohn, and Nirenberg (1982).

For the 2-dimensional Navier-Stokes equations (2DNS), it is shown that for any F satisfying (1.4) and any $U_0 \in H$, there exists a unique weak solution

$$U(t) \in L^\infty(0, \infty; H) \cap L^2(0, \tau; V) \cap C_w^0([0, \infty); H)$$

for any $\tau > 0$, see Lions and Prodi (1959). Furthermore, if $U_0 \in V$, there exists a unique global strong solution

$$U(t) \in L^\infty(0, \infty; V) \cap C^0([0, \infty); V) \cap L^2(0, \tau; D(A))$$

for any $\tau > 0$, see Ladyzhenskaya (1967, 1969, 1972); and there exists a positive constant L_1, independent of U_0 such that

$$(1.8) \qquad \limsup_{t \to \infty} \|U(t)\|_V \leq L_1.$$

If, in addition $F(\cdot)$ belongs to $W^{1,\infty}([0,\infty), L^2(\Omega))$ for instance, the strong solution $U(t)$ is in $C^0((0,\infty); D(A))$ and there exists a positive constant L_2, independent of U_0 such that

(1.9) $$\limsup_{t\to\infty}\|AU(t)\|_H \le L_2.$$

All these classical results (and more precise ones) can be found in Ladyzhenskaya (1969), Temam (1977, 1983, 1988) as well as in Constantin and Foias (1988).

As a result of (1.8) and (1.9), it follows that in 2-dimensions, when F is time independent, one can define a C^0-semigroup $S(t)$ acting on V (or also on H). This semigroup has a global attractor $\mathcal{A}$ in V, that is to say, there exists a compact, invariant set $\mathcal{A}$ that attracts any bounded set B in V, see Ladyzhenskaya (1972). We recall that $\mathcal{A}$ is invariant if $S(t)\mathcal{A} = \mathcal{A}$ for all $t \ge 0$ and that $\mathcal{A}$ attracts any bounded set B in V if, for any positive number ε and any bounded set B in V, there exists a positive time $\tau = \tau(\varepsilon, B)$ such that, for $t \ge \tau$, the set $T(t)B$ is contained in the ε-neighbourhood $\mathcal{N}_V(\mathcal{A}, \varepsilon)$ of $\mathcal{A}$ in V. Actually $\mathcal{A}$ is compact in $H^2(\Omega)$ and is the global attractor of the 2DNS equations in $H^2(\Omega) \cap V$. The global attractor $\mathcal{A}$ has also finite dimension, see Mallet-Paret (1976). The dimensionality of $\mathcal{A}$ has been widely studied (see, for instance, Foias and Temam (1979), Babin and Vishik (1983), Constantin and Foias (1985), Constantin, Foias and Temam (1985)). If F is time-varying, but has some compactness property, one can still show that (1.3) has a global attractor in $V \times \mathcal{F}$, where $\mathcal{F}$ is a compact, positively invariant subset of $W^{1,\infty}([0,\infty), L^2(\Omega))$.

For the 3-dimensional Navier-Stokes equations (3DNS) the situation is quite different. It is known that, for any F satisfying (1.4) and any $U_0 \in H$, there exists at least a weak solution

$$U(t) \in L^\infty(0,\infty; H) \cap L^2(0,\tau,V) \cap C^0_w([0,\infty); H),$$

and $\frac{dU}{dt} \in L^{4/3}_{loc}(0,\tau; V')$ for every $\tau > 0$. Furthermore, if U_0 belongs to V, there exists a T, which depends on F and U_0, $0 < T \le \infty$, such that (1.3) has a unique strong, or regular, solution

$$U(t) \in L^\infty(0,T;V) \cap C^0([0,T];V) \cap L^2(0,\tau; D(A))$$

and $\frac{dU}{dt} \in L^2(0,\tau; H)$ for every finite number $\tau, 0 < \tau \le T$. Moreover, if the data $U_0 \in V$ and F are small, then one has $T = \infty$, i.e., (1.3) has a global regular solution for small data. For generic solvability of the 3DNS equations, one refers to Fursikov (1980). For other results, see Ladyzhenskaya (1969), Temam (1977, 1983, 1988) and Constantin and Foias (1988).

When F is time independent, by lack of uniqueness of the weak solutions of (1.3), one can no longer define a semigroup $S(t)$ on H. However Foias and Temam (1987) have constructed a set $\mathcal{X}$ compact in H_{weak}, that we call weak global attractor of (1.3). Indeed, let $\mathcal{X}$ denote the set of those $U_0 \in H$ for which there exists a weak solution $U(t)$ of (1.3) on $(-\infty, \infty)$ that is bounded in H on $(-\infty, +\infty)$, such that

$U(0) = U_0$. The set $\mathcal{X}$ is not empty since (1.3) has stationary solutions and $\mathcal{X}$ contains these solutions. Moreover, $\mathcal{X}$ is invariant in some sense, and $\mathcal{X}$ is compact in H_{weak}. Furthermore, for every weak solution $U(t)$ of (1.3) in $(0, \infty)$, we have

$$(1.10) \qquad\qquad U(t) \to \mathcal{X} \text{ in } H_{\text{weak}} \text{ as } t \to +\infty.$$

A first question arises:

$$(1.11) \qquad\qquad \text{Is } \mathcal{X} \text{ contained in } V \text{ and bounded in } V?$$

We will answer this question in the case of the thin 3-dimensional domains Ω_ε.

It is the existence of the global attractor for the 2DNS which is the *raison d'être* of our study of the 3DNS on thin domains. Because a thin 3-dimensional domain is somehow *close* to a 2-dimensional domain, it is natural to wonder whether one can use the good global properties of the 2DNS to study the global regularity of the 3DNS. In Raugel and Sell (1990a), we gave a positive answer to this question.

The idea of exploiting the existence of a global attractor of an evolutionary equation on a n-dimensional domain to obtain better global properties for a corresponding equation on a thin $(n + 1)$-dimensional domain has been first used in Hale and Raugel (1989a), (1989b) (see also Raugel (1988)). Such results have been extended to general thin $(n + k)$-dimensional domains and various evolutionary equations in Hale and Raugel (1990a), (1990b), (1991).

The fact that Ω_ε is close to Q_2 implies that the 3DNS on Ω_ε is a perturbation of the 2DNS on Q_2, but this perturbation is singular. To overcome this difficulty, we follow the methods introduced in Hale and Raugel (1989a, 1989b) for reaction diffusion equations and damped wave equations on more general thin domains. Let us describe quickly these methods for our particular thin domain $\Omega_\varepsilon = Q_2 \times (0, \varepsilon)$, in the case of periodicity in the third variable y_3. First one maps Ω_ε onto $Q_3 = Q_2 \times (0, 1)$ by means of dilation. The Navier-Stokes equations (1.1) on Ω_ε are then transformed to the dilated Navier-Stokes equations on Q_3, (see (2.10) below and the abstract form (2.13)). The differential operators appearing in (2.10) or (2.13) are singular in the sense that they contain coefficients with ε^{-1} or ε^{-2}, where ε is small. The second step is accomplished by introducing the orthogonal projection $v = Mu$ where

$$(1.12) \qquad\qquad v(y_1, y_2) = \frac{1}{\varepsilon} \int_0^\varepsilon u(y_1, y_2, s)\,ds$$

By applying M and $(I - M)$ to the dilated Navier-Stokes evolutionary equation (2.13), one finds an equivalent system (2.23) in v and $w \equiv u - v$. Then using some more accurate Sobolev inequalities given in Hale and Raugel (1989b), Poincaré inequalities and several a priori estimates, we show that the system (2.23) is actually a regular perturbation of the 2DNS, when ε is small, in the sense that the system (2.23) is a regular perturbation of the reduced 3D Navier-Stokes evolutionary equation (2.24). In Raugel and Sell (1990a), we have described the above method in details and we have proved the following result.

THEOREM A. *Consider the 3DNS (1.1) on Ω_ε with periodic boundary conditions in the third variable (i.e., $U|_{y_3=0} = U|_{y_3=\varepsilon}$). There exists a positive number $\varepsilon_0 = \varepsilon_0(\nu)$ such that, for every ε, $0 < \varepsilon \leq \varepsilon_0$, there are large sets $\mathcal{R}(\varepsilon)$ and $\mathcal{S}(\varepsilon)$ with*

$$\mathcal{R}(\varepsilon) \subset V, \quad \mathcal{S}(\varepsilon) \subset L^\infty(0, \infty, L^2(\Omega_\varepsilon))$$

such that, if $U_0 \in \mathcal{R}(\varepsilon)$ and $F \in \mathcal{S}(\varepsilon)$, then (1.3) has a (unique) global strong solution

$$U(t) \in L^2(0, \tau; D(A)) \cap L^\infty(0, \infty; V) \cap C^0([0, \infty); V)$$

for any $\tau > 0$. Furthermore there exists a positive constant L_1 (which depends on F and not of U_0) such that

$$\limsup_{t \to \infty} \|U(t)\|_V \leq L_1.$$

In Section 2, we will clarify the assertion that $\mathcal{R}(\varepsilon)$ and $\mathcal{S}(\varepsilon)$ are large sets. In particular, we will recall the precise statement given in Raugel and Sell (1990a), (see Theorem 1 of Section 2). As ε goes to zero, the diameters of $\mathcal{R}(\varepsilon)$ and $\mathcal{S}(\varepsilon)$ go to infinity. In the case of periodic boundary conditions in all the variables, one can show that the sets $\mathcal{R}(\varepsilon)$ and $\mathcal{S}(\varepsilon)$ can be replaced by larger sets $\mathcal{R}^*(\varepsilon)$ and $\mathcal{S}^*(\varepsilon)$ (see Raugel and Sell (1990b)). This improvement is given in Theorem 2 of Section 2.

Assume that F belongs to $\mathcal{S}(\varepsilon)$ and is time-independent. It is a consequence of the above regularity results that one can define a C^0 semigroup $S_\varepsilon(t)$ on $\mathcal{R}(\varepsilon)$. Moreover, the set of strong solutions of the Navier-Stokes evolutionary equation (1.3) has a local attractor $\mathcal{A}_\varepsilon \equiv \mathcal{A}_\varepsilon(F)$ attracting every bounded set contained in $\bigcup_{t \geq 0} S_\varepsilon(t)\mathcal{R}(\varepsilon)$. This attractor $\mathcal{A}_\varepsilon$ is compact in $H^2(\Omega_\varepsilon) \cap V$.

In Section 3, we prove the following result.

THEOREM B. *Under the hypotheses of Theorem A, the local attractor $\mathcal{A}_\varepsilon \equiv \mathcal{A}_\varepsilon(F)$ is actually the global attractor for all the weak solutions of the Navier-Stokes evolutionary equation (1.3) provided that F belongs to $\mathcal{S}(\varepsilon)$. In particular, $\mathcal{A}_\varepsilon$ coincides with the weak attractor $\mathcal{X}_\varepsilon$ of (1.3).*

Since $\mathcal{A}_\varepsilon$ and $\mathcal{X}_\varepsilon$ coincide and that $\mathcal{A}_\varepsilon \subset V$, we have a positive answer to the question (1.11), in our case. If F is time-varying but satisfies some compactness property and belongs to $\mathcal{S}(\varepsilon)$, we can still show an analog of Theorem B.

In the next section, we shall introduce the notation, the projection operator M and the v and w equations. There we shall also state the regularity theorems given in Raugel and Sell (1990a, 1990b). In Section 3, we shall construct the local attractor $\mathcal{A}_\varepsilon$, recall the proof of Theorem B, and state upper semicontinuity results.

2. Notation-statement of the regularity theorems.

We recall that the Navier-Stokes equations on a bounded domain Ω in $\mathbf{R}^n$, $n = 2, 3$, are given by

$$(2.1) \qquad U_t - \nu \Delta U + (U \cdot \nabla)U + \nabla P = F, \quad \nabla \cdot U = 0.$$

Here we are interested in the case $n = 3$, when $\Omega \equiv \Omega_\varepsilon$ is a thin 3-dimensional domain, i.e., $\Omega_\varepsilon = Q_2 \times (0, \varepsilon)$, where ε is a small positive parameter $0 < \varepsilon \leq 1$, $Q_2 = (0, \ell_1) \times (0, \ell_2)$, $\ell_i > 0$, $i = 1, 2$, or Q_2 is a smooth bounded domain in $\mathbf{R}^2$. The equation (2.1) can be supplemented with various boundary conditions. When $Q_2 = (0, \ell_1) \times (0, \ell_2)$, we assume that $\varepsilon \leq \ell_2 \leq \ell_1$ and (2.1) is supplemented with the periodic and zero mean conditions

(2.2)
$$\begin{cases} \text{(i)} \quad U(y + \ell_i e_i, t) = U(y, t), i = 1, 2, U(y + \varepsilon e_3, t) = U(y, t), \\ \text{and} \\ \text{(ii)} \quad \int_{\Omega_\varepsilon} U \, dy = 0 \end{cases}$$

where $\{e_1, e_2, e_3\}$ is the natural basis in $\mathbf{R}^3$. In this case, we require, of course, that F and the initial data U_0 satisfy

$$\int_{\Omega_\varepsilon} F \, dy = \int_{\Omega_\varepsilon} U_0 \, dy = 0.$$

It then follows that any solution U of (2.1) with $U(0) = U_0$ will also satisfy $\int_{\Omega_\varepsilon} U \, dy = 0$ for $t > 0$. To simplify the notation, we set: $\ell_3 = 1$. In the case where Q_2 is a *smooth* bounded domain in $\mathbf{R}^2$, we are interested by the following boundary conditions, here called *mixed conditions*:

(2.3)
$$\begin{cases} \text{(i)} \quad U(y + \varepsilon e_3, t) = U(y, t), \\ \text{and} \\ \text{(ii)} \quad U(y, t) = 0 \text{ on } \Gamma_{2\varepsilon} = \partial Q_2 \times (0, \varepsilon). \end{cases}$$

We set:
$$\Gamma_{1\varepsilon} = Q_2 \times \{0\} \cup Q_2 \times \{\varepsilon\}$$

At the end of Section 3, we will make some comments about the case of homogeneous Dirichlet boundary conditions

(2.4)
$$U(y, t) = 0 \text{ on } \partial \Omega_\varepsilon.$$

In all cases, we set: $Q_3 = Q_2 \times (0, 1)$. The change of variables $(y_1, y_2, y_3) \mapsto (x_1, x_2, x_3)$ where

(2.5)
$$x_i = y_i, \quad i = 1, 2, \quad \text{and } x_3 = \varepsilon^{-1} y_3,$$

maps Ω_ε onto Q_3. We set $\Gamma_1 = Q_2 \times \{0\} \cup Q_2 \times \{1\}$ and $\Gamma_2 = \partial Q_2 \times (0, 1)$.

2.1 Dilated Navier-Stokes equations. The linear operator J_ε given by $U = J_\varepsilon u$, where

(2.6)
$$U(y_1, y_2, y_3) = u(y_1, y_2, \varepsilon^{-1} y_3),$$

sets up a one-to-one correspondence between measurable functions on Ω_ε and measurable functions on Q_3. The following identities are easily verified,

$$(2.7) \qquad \|J_\varepsilon u\|^p_{L^p(\Omega_\varepsilon)} = \varepsilon \|u\|^p_{L^p(Q_3)}, \quad 1 \le p < \infty$$

and

$$(2.8) \qquad \frac{\partial}{\partial y_i} J_\varepsilon u = J_\varepsilon \varepsilon^{-\{i\}} \frac{\partial}{\partial x_i} u, \quad i = 1, 2, 3, \text{ where } \{1\} = \{2\} = 0, \{3\} = 1.$$

From (2.7) and (2.8), we deduce that

$$(2.9) \qquad \left\|\frac{\partial}{\partial y_i} J_\varepsilon u\right\|^p_{L^p(\Omega_\varepsilon)} = \varepsilon \left\|\varepsilon^{-\{i\}} \frac{\partial}{\partial x_i} u\right\|^p_{L^p(Q_3)}, u \in W^{1,p}(Q_3), \quad 1 \le p < \infty.$$

We see that $J_\varepsilon(W^{k,p}(Q_3)) = W^{k,p}(\Omega_\varepsilon)$ for any Sobolev space $W^{k,p}$. Henceforth we shall use capital Roman letters to denote functions on Ω_ε and lower case Roman letters for functions on Q_3.

We want to let ε vary in our study of the solutions of (2.1). Rather than studying a *fixed equation on a variable domain*, it is more convenient to *fix the domain and to let the equation vary*. Here we follow the method of Hale and Raugel (1989a). In particular, by using the operator J_ε, the Navier-Stokes equations (2.1) with the boundary conditions (2.2) or (2.3) or (2.4) are transformed into the following system on Q_3:

$$(2.10) \qquad \begin{cases} u_t - \nu \Delta_\varepsilon u + (u \cdot \nabla_\varepsilon) u + \nabla_\varepsilon p = f \\ \nabla_\varepsilon \cdot u = 0 \end{cases}$$

where $\nabla_\varepsilon = (D_1, D_2, \varepsilon^{-1} D_3), \Delta_\varepsilon = D_1^2 + D_2^2 + \varepsilon^{-2} D_3^2, D_i = \frac{\partial}{\partial x_i}, i = 1, 2, 3, u = J_\varepsilon^{-1} U, f = J_\varepsilon^{-1} F, p = J_\varepsilon^{-1} P$. Of course, the system (2.10) is supplemented with boundary conditions corresponding to (2.2), (2.3) or (2.4) that will be described in the next section. We will refer to (2.10) as *the dilated Navier-Stokes equations* on Q_3.

2.2 Abstract formulation. The next step is to formulate the initial value problem for (2.10) supplemented with boundary conditions as an abstract nonlinear evolutionary equation on an adequate Hilbert space H_ε.

Let $L^2(Q_3) = L^2(Q_3)^3$ be the set (of the classes) of functions $u : Q_3 \to \mathbf{R}^3$ which are square-integrable and let us denote by $\|\cdot\|$ its usual norm. For $m = 0, 1, 2, \ldots$ we denote by $H^m(Q_3)$ the Sobolev space $H^m(Q_3)^3$. If $Q_2 = (0, \ell_1) \times (0, \ell_2)$, we denote by $H_p^m(Q_3) = H_p^m(Q_3)^3$ the closure in $H^m(Q_3)$ of those smooth functions which are periodic in space, i.e., $u(x + \ell_i e_i) = u(x), i = 1, 2, 3$. Of course, $H_p^0(Q_3) = L^2(Q_3)$.

We now define the space H_ε and V_ε which depend on the boundary conditions:

- if the equations (2.1) are supplemented with the conditions (2.2), we denote by $H_\varepsilon = H_\varepsilon(Q_3)$ (respectively by $V_\varepsilon = V_\varepsilon(Q_3)$) the closure in $L^2(Q_3)$ (respectively $H^1(Q_3)$) of those smooth functions u that are periodic on Q_3 and satisfy

$$(2.11) \qquad \int_{Q_3} u\, dx = 0 \text{ and } \operatorname{div}_\varepsilon u = \nabla_\varepsilon \cdot u = D_1 u_1 + D_2 u_2 + \varepsilon^{-1} D_3 u_3 = 0.$$

– if the equations (2.1) are supplemented with the conditions (2.3), we denote by $H_\varepsilon = H_\varepsilon(Q_3)$ (resp. $V_\varepsilon = V_\varepsilon(Q_3)$) the closure in $\mathbf{L}^2(Q_3)$ (resp. $\mathbf{H}^1(Q_3)$) of those smooth functions u that satisfy periodic boundary conditions on Γ_1 and homogeneous Dirichlet boundary conditions on Γ_2 and $\nabla_\varepsilon \cdot u = 0$ in Q_3.

– finally, if the equations (2.1) are supplemented with the boundary conditions (2.4), we denote by H_ε (resp. V_ε) the closure in $\mathbf{L}^2(Q_3)$ (resp. $\mathbf{H}^1(Q_3)$) of those smooth functions u that satisfy homogeneous Dirichlet boundary conditions on ∂Q_3 and $\nabla_\varepsilon \cdot u = 0$ in Q_3.

In all three cases, we let $\mathbf{P}_\varepsilon$ denote the orthogonal projection of $\mathbf{L}^2(Q_3)$ onto H_ε. Introducing the operators

$$(2.12) \qquad A_\varepsilon u = -\mathbf{P}_\varepsilon \Delta_\varepsilon u, \quad B_\varepsilon(u^1, u^2) = \mathbf{P}_\varepsilon(u^1 \cdot \nabla_\varepsilon)u^2,$$

and applying the operator $\mathbf{P}_\varepsilon$ to (2.10), we obtain the following abstract nonlinear evolutionary equation on H_ε

$$(2.13) \qquad u' + \nu A_\varepsilon u + B_\varepsilon(u, u) = \mathbf{P}_\varepsilon f.$$

We refer to (2.13) as the *dilated Navier-Stokes evolutionary equation*. We define V_ε^m for $m = 1, 2, \ldots$ by

$$V_\varepsilon = V_\varepsilon^m = V_\varepsilon \cap \mathbf{H}_p^m(Q_3) \quad (\text{resp. } V_\varepsilon^m = V_\varepsilon \cap \mathbf{H}^m(Q_3))$$

in the case of the boundary conditions (2.2) (resp. for the other boundary conditions). One has:

$$V_\varepsilon^1 = D(A_\varepsilon^{1/2}) \text{ and } D(A_\varepsilon) = V_\varepsilon^2.$$

The proof of the second equality above is easy when we have the boundary conditions (2.2) or (2.3). In the case of homogeneous Dirichlet boundary conditions, the second equality is a consequence of regularity results (see Dauge (1984, 1989) for instance). Let us recall that, in the periodic case, we have:

$$(2.14) \qquad A_\varepsilon u = -\Delta_\varepsilon u, \quad u \in D(A_\varepsilon).$$

Also, in the periodic case, the spaces V_ε^m can be described in terms of the Fourier series expansion for functions $u \in \mathbf{L}^2(Q_3)$. Likewise, $A_\varepsilon u$ can be represented in terms of the Fourier series expansion of u (See Raugel and Sell (1990a)).

Using the classical Poincaré inequality, one easily shows that there exist two positive constants C_1 and C_2, independent of ε, such that,

$$(2.15) \qquad \begin{aligned} C_1(\|u\|_{\mathbf{H}^1(Q_3)} + \varepsilon^{-1}\|D_3 u\|_{\mathbf{L}^2(Q_3)}) &\leq \|A_\varepsilon^{1/2} u\|_{\mathbf{L}^2(Q_3)} \\ &\leq C_2(\|u\|_{\mathbf{H}^1(Q_3)} + \varepsilon^{-1}\|D_3 u\|_{\mathbf{L}^2(Q_3)}). \end{aligned}$$

In the case of the boundary conditions (2.2) and (2.3), one also has (see Raugel, Sell (1990a)):

$$(2.16) \qquad \begin{aligned} C_1(\|u\|_{\mathbf{H}^2(Q_3)} + \varepsilon^{-1}\|D_3 u\|_{\mathbf{H}^1(Q_3)} &+ \varepsilon^{-2}\|D_3^2 u\|_{\mathbf{L}^2(Q_3)}) \\ &\leq \|A_\varepsilon u\|_{\mathbf{L}^2(Q_3)} \\ &\leq C_2(\|u\|_{\mathbf{H}^2(Q_3)} + \varepsilon^{-1}\|D_3 u\|_{\mathbf{H}^1(Q_3)} + \varepsilon^{-2}\|D_3^2 u\|_{\mathbf{L}^2(Q_3)}) \end{aligned}$$

We will assume the forcing term f in (2.10) to be a time-varying function in the space $L^\infty((0,\infty); L^2(Q_3))$ and we define the norm $\|f\|_\infty$ by

$$\|f\|_\infty \equiv \operatorname*{ess\,sup}_{0<t<\infty} \|f(t)\|.$$

Finally, we define the trilinear form $b_\varepsilon(u,v,w)$ by

$$(2.17) \qquad b_\varepsilon(u,v,w) = \sum_{i,j=1}^{3} \int_{Q_3} \varepsilon^{-\{i\}} u_i(D_i v_j) w_j dx.$$

Clearly, we have:

$$\langle B_\varepsilon(u,v), w\rangle = b_\varepsilon(u,v,w), \quad \text{if } u,v,w \in V_\varepsilon^1.$$

From now on and till the end of Section 3, we will only consider the boundary conditions (2.2) or (2.3).

2.3 The projection M and the v and w equations. We want to compare $u \in L^2(Q_3)$ with an element $v \in L^2(Q_2)$. To this end, we introduce the orthogonal projection M on $L^2(Q_3)$ defined by $v = Mu$ with

$$(2.18) \qquad (Mu)(x_1, x_2) = v(x_1, x_2) = \int_0^1 u(x_1, x_2, s) ds$$

and we set $w = (I - M)u$. Since $w = (I - M)u$, one has $Mw = 0$ and it is clear that M is an orthogonal projection on $L^2(Q_3)$ which satisfies

$$MD_i u = D_i Mu, i = 1,2, \text{ for all } u \in W^{1,1}(Q_3)^3,$$
$$MD_3 u = D_3 Mu = 0, \text{ for all } u \in W^{1,1}(Q_3)^3 \text{ such that } u(x + \ell_3 e_3) = u(x).$$

Therefore, for these functions u, we have:

$$(2.19) \qquad \nabla_\varepsilon \cdot Mu = M\nabla_\varepsilon \cdot u.$$

Using the above equalities, one also shows that, for all u in $D(A_\varepsilon)$,

$$(2.20) \qquad \begin{cases} MA_\varepsilon u = MA_\varepsilon Mu = A_\varepsilon Mu, \\ (I - M)A_\varepsilon u = A_\varepsilon(I - M)u. \end{cases}$$

In the periodic case, (2.20) is an obvious consequence of the above identities and of (2.14). In the case of the boundary conditions (2.3), we remark that (2.20) holds if and only if

$$(2.21) \qquad M\mathbf{P}_\varepsilon = \mathbf{P}_\varepsilon M, \quad \text{on } H_\varepsilon.$$

Let $u = Mu + (I - M)u = v + w$ be any element of $\mathbf{L}^2(Q_3)$. The equality (2.21) is proved if we show that $MP_\varepsilon w = 0$ and $MP_\varepsilon v = P_\varepsilon v$. Since M and P_ε are orthogonal projections in $\mathbf{L}^2(Q_3)$ and $\operatorname{div}_\varepsilon MP_\varepsilon w = 0$, we have:

$$\int_{Q_3} (MP_\varepsilon w)^2 dx_1 dx_2 dx_3 = \int_{Q_3} (MP_\varepsilon w)(P_\varepsilon w) dx_1 dx_2 dx_3 = \int_{Q_3} (MP_\varepsilon w) w \, dx_1 dx_2 dx_3 = 0,$$

which implies that $MP_\varepsilon w = 0$. Likewise one shows that $MP_\varepsilon v = P_\varepsilon v$.

In Hale and Raugel (1989a) and in Raugel and Sell (1990a), the following Poincaré inequality is proved,

$$(2.22) \qquad (i) \qquad \|w\| \leq C_3 \varepsilon \|A_\varepsilon^{1/2} w\|, \quad w \in V_\varepsilon, Mw = 0,$$

where c_3 is a positive constant independent of ε. Furthermore in Hale and Raugel (1989b) (see also Raugel and Sell (1990a, Appendix)), it is also shown that,

$$(2.22) \qquad (ii) \qquad \|w\|_{L^q(Q_3)^3} \leq C_3 \varepsilon^{2/q} \|A_\varepsilon^{1/2} w\|, w \in V_\varepsilon, Mw = 0,$$

for $2 \leq q \leq 6$, which leads to more accurate estimates of $|b_\varepsilon(u^1, u^2, u^3)|$ when u^1, u^2 or u^3 satisfy the above conditions (Raugel and Sell (1990a, Appendix)).

We now apply the projections M and $(I - M)$ to the equation (2.13) where $v = Mu$ and $w = (I - M)u$. Since one has $MB_\varepsilon(v, v) = B_\varepsilon(v, v)$, it follows from (2.20) that one obtains the system:

$$(2.23) \quad \begin{cases} v' + \nu A_\varepsilon v + B_\varepsilon(v, v) = MP_\varepsilon f - M(B_\varepsilon(v, w) + B_\varepsilon(w, v) + B_\varepsilon(w, w)) \\ w' + \nu A_\varepsilon w = (I - M)P_\varepsilon f - (I - M)(B_\varepsilon(v, w) + B_\varepsilon(w, v) + B_\varepsilon(w, w)). \end{cases}$$

The initial condition $u(0) = u_0 = v_0 + w_0$ also splits into a v and w component. Note that $A_\varepsilon v$ and $B_\varepsilon(v, v)$ are independent of ε (In the periodic case, one has $A_\varepsilon v = D_1^2 v + D_2^2 v$). We are going to study the solutions $(v, w) = (v(t), w(t))$ of (2.23) such that $v(0) = v_0 \equiv Mu_0$, $w(0) = w_0 = (I - M)u_0$.

2.4 Reduced 3D Navier-Stokes evolutionary equation. When

$$(I - M)P_\varepsilon f = 0,$$

then the set $\{w = 0\}$ is positively invariant, that is, if $u_0 = v_0 + w_0$ depends only on x_1 and x_2 (i.e., $w_0 = 0$), then $w(t) \equiv 0$ for all $t \geq 0$ and $\bar{v} = v(t)$ satisfies the equation

$$(2.24) \qquad \qquad \bar{v}' + \nu A_\varepsilon \bar{v} + B_\varepsilon(\bar{v}, \bar{v}) = MP_\varepsilon f,$$

with $\bar{v}(0) = v_0$. We refer to (2.24) as *the reduced 3D Navier-Stokes evolutionary equation*. Note that $\bar{v} = (\bar{v}_1, \bar{v}_2, \bar{v}_3)$ is a three dimensional vector field on Q_3, independent of x_3. The reduced 3D Navier-Stokes evolutionary equation somehow incorporates the 2DNS equations on Q_2. In order to see this, we let $L^2(Q_2)^2$

denote the L^2-space of 2-dimensional vector fields $m = (m_1, m_2)$ which depend on $(x_1, x_2) \in Q_2$. In the case of the boundary conditions (2.2) (respectively (2.3)), we let $H(Q_2)$ denote the closure in $L^2(Q_2)^2$ of those smooth functions m that are periodic on Q_2, and satisfy $\int_{Q_2} m \, dx = 0$ and $D_1 m_1 + D_2 m_2 = 0$ (respectively, of those smooth functions m that satisfy homogeneous Dirichlet boundary conditions on ∂Q_2 and $D_1 m_1 + D_2 m_2 = 0$). Finally we let $\mathbf{P}_2$ denote the orthogonal projection of $L^2(Q_2)^2$ onto $H(Q_2)$. Clearly one has,

$$(2.25) \qquad \mathbf{P}_\varepsilon \begin{pmatrix} g_1 \\ g_2 \\ g_3 \end{pmatrix} = \begin{pmatrix} \mathbf{P}_2 \begin{pmatrix} g_1 \\ g_2 \end{pmatrix} \\ g_3 \end{pmatrix}, \qquad \text{for all } g = (g_1, g_2, g_3) \in M(L^2(Q_2)^2).$$

Furthermore one checks at once that $\bar{v}$ is a solution of the reduced 3D Navier-Stokes evolutionary equation (2.24) if and only if $m = (\bar{v}_1, \bar{v}_2)$ is a solution of the 2D Navier-Stokes evolutionary equation

$$(2.26) \qquad \text{(i)} \qquad \frac{dm}{dt} - \nu \mathbf{P}_2 (D_1^2 + D_2^2) m + \mathbf{P}_2 (m \cdot \nabla) m = (g_1, g_2)$$

and $\bar{v}_3$ is a solution of the linear equation

$$(2.26) \qquad \text{(ii)} \qquad \frac{d\bar{v}_3}{dt} - \nu (D_1^2 + D_2^2) \bar{v}_3 + (\bar{v}_1 D_1 + \bar{v}_2 D_2) \bar{v}_3 = g_3,$$

where $g = (g_1, g_2, g_3) = M \mathbf{P}_\varepsilon f$.

2.5 Statement of the regularity results. Here we shall give a precise description of the sets $\mathcal{R}(\varepsilon)$ and $\mathcal{S}(\varepsilon)$ given in Theorem A. To this end, we introduce the following hypothesis $H(a, b)$. We shall say that the bounded monotone functions $\eta_i(\varepsilon)$ defined for $0 < \varepsilon \leq 1$, $i = 1, 2, 3, 4$, and the constants r and p satisfy Hypothesis $H(a, b)$, where a and b are positive, provided:

(1) $p \geq -1, r > -2$,
(2) $\varepsilon^{1/4} \eta_i^{-1} \to 0$ as $\varepsilon \to 0, i = 1, 2$,
(3) $\varepsilon^{1/8} \eta_i^{-1} \to 0$ as $\varepsilon \to 0, i = 3, 4$,
(4) $\varepsilon^{1/4} Q(\varepsilon)$ is bounded for $0 < \varepsilon \leq 1$, where

$$Q(\varepsilon) = |\log(2 C_3^2 \nu^{-2} \varepsilon^{2 + r - p} \eta_4^{-2} \eta_3^2)|,$$

(5) Let $a > 0$ be fixed, then one has

$$(2.27) \qquad \begin{cases} \varepsilon^{5/8} \eta^{-2} \exp(a \eta^{-4}) \to 0 \\ \eta^{-2} \to 0 \qquad \text{as } \varepsilon \to 0^+ \end{cases}$$

and $\varepsilon^{5/8} \exp(2 a \eta_2^{-4})$ is bounded for $0 < \varepsilon \leq 1$, where

$$(2.28) \qquad \eta^{-2} \equiv \max(4 \eta_1^{-2} + k_1^2 \eta_3^{-4} + k_2^2 \varepsilon^{2 + r} \eta_4^{-2}, 1)$$

and k_1, k_2 are two adequate positive constants depending only on ν and Q_2,

(6) Let $b > 0$ be fixed; then, for any $\lambda, 0 < \lambda < 1$, there is a positive number $\varepsilon_4 = \varepsilon_4(b, \lambda) > 0$ such that one has

$$(2.29) \qquad \eta_2^{-2} \exp(b\eta_2^{-4}) \leq \lambda(4\eta_1^{-2} + k_1^2\eta_3^{-4}), \quad 0 < \varepsilon \leq \varepsilon_4,$$

and

(7) the function $\varepsilon^{4+2r}\eta_4^{-4}(\log \eta^{-4} + 1)$ is bounded as $\varepsilon \to 0^+$.

In Raugel and Sell (1990a), we gave an example where Hypothesis $H(a,b)$ holds. Let us recall it.

Example 1. We fix $p \geq -1$ and set

$$\begin{cases} r & = -2 + \delta, \quad \delta > 0 \\ \eta_4^{-1} & = -\log \varepsilon \\ \eta_3^{-2} & = \eta_1^{-1} \end{cases}$$

and we define:

$$\eta_1^{-2} = (-\log \varepsilon^\alpha)^{1/2}, \quad \eta_2^{-2} = (\log(-\log \varepsilon^\alpha)^\beta)^{1/2},$$

where $\alpha > 0$, $(48 + 3k_1^4)a\alpha < \frac{5}{8}$, $\beta > 0, 2b\beta < 1$.

In Raugel and Sell (1990a), we have proved the following result.

THEOREM 1. *Assume that the Navier-Stokes equations (2.1) are supplemented with the boundary conditions (2.2) or (2.3). Let $\eta_i, i = 1,2,3,4,r$ and p satisfy Hypothesis $H(a,b)$, where a and b are sufficiently large. Then there exist a positive number ε_0, a continuous function $\Gamma \in C^0([0,\infty), \mathbf{R})$, and for all $\varepsilon, 0 < \varepsilon \leq \varepsilon_0$, there exists $\widehat{T}_1 = \widehat{T}_1(\varepsilon) > 0$, such that, whenever $u_0 \in D(A_\varepsilon^{1/2}), f \in L^\infty((0,\infty), \mathbf{L}^2(Q_3))$ satisfy*

$$(2.30) \qquad \begin{cases} \|A_\varepsilon^{1/2}v_0\|^2 \leq \eta_1^{-2}, & \|M\mathbf{P}_\varepsilon f\|_\infty^2 \leq \eta_2^{-2}, \\ \|A_\varepsilon^{1/2}w_0\|^2 \leq \varepsilon^p\eta_3^{-2}, & \|(I-M)\mathbf{P}_\varepsilon f\|_\infty^2 \leq \varepsilon^r\eta_4^{-2}, \end{cases}$$

then (2.13) has a (unique) solution u that belongs to $C^0([0,\infty), V_\varepsilon) \cap L^\infty([0,\infty), V_\varepsilon)$, i.e., one has

$$(2.31) \qquad \|A_\varepsilon^{1/2}u(t)\|^2 \leq K_1^2, \quad t \geq 0,$$

where K_1 depends only on $\nu, \eta_i, i = 1,2,3,4$ and Q_2. Moreover, the components of $u = v + w$ satisfy

$$(2.32) \qquad \|A_\varepsilon^{1/2}v(t)\|^2 \leq \Gamma(\eta_2^{-2}), \quad t \geq \widehat{T}_1,$$

and

$$(2.33) \qquad \|A_\varepsilon^{1/2}w(t)\|^2 \leq k_2^2\varepsilon^{2+r}\eta_4^{-2}, \quad t \geq \widehat{T}_1$$

Remark 2.1. If, in addition to the hypotheses of Theorem 1, $\mathbf{P}_\varepsilon f$ belongs to $C^0([0,\infty), H_\varepsilon) \cap L^\infty((0,\infty), H_\varepsilon) \cap W^{1,\infty}((0,\infty), D(A_\varepsilon^{-1/2}))$ then the solution u of (2.13) belongs to $C^0((0,\infty), V_\varepsilon^2) \cap L^\infty([t_0,\infty), V_\varepsilon^2)$ for any $t_0 > 0$. If, in addition u_0 belongs to V_ε^2, the solution u of (2.13) is in $C^0([0,\infty), V_\varepsilon^2)$.

Remark 2.2. The positive constants k_1, k_2 appearing in the hypothesis $H(a, b)$ and in Theorem 1, as well as the function Γ given in Theorem 1, are determined in Raugel and Sell (1990a), Section 3.

Remark 2.3. Theorem 1 is proved in two steps in Raugel and Sell (1990a). The first step consists in showing that if u_0 and f satisfy (2.30), there is a positive time $T_1 = T_1(\varepsilon)$ such that $u(t) \in D(A_\varepsilon^{1/2})$ for $0 \le t \le T_1$ and

$$(2.34) \qquad \begin{cases} \|A_\varepsilon^{1/2} v(T_1)\|^2 \le 4\eta_1^{-2} + k_1^2 \eta_3^{-4} \\ \|A_\varepsilon^{1/2} w(T_1)\|^2 \le k_2^2 \varepsilon^{2+r} \eta_4^{-2}. \end{cases}$$

The second step consists in showing that if u_0 and f satisfy (2.34) and (2.30) respectively, then there is a positive time $T_0 = T_0(\varepsilon)$ such that $u(t) \in D(A_\varepsilon^{1/2})$ for $0 \le t \le 2T_0$ and

$$(2.35) \qquad \begin{cases} \|A_\varepsilon^{1/2} v(t)\|^2 \le \frac{1}{2}(4\eta_1^{-2} + k_1^2 \eta_3^{-4}) \\ \|A_\varepsilon^{1/2} w(t)\|^2 \le k_2^2 \varepsilon^{2+r} \eta_4^{-1}, \end{cases}$$

for $T_0 \le t \le 2T_0$.

Below we will explain that the conditions (2.30) correspond to "large data" given on Ω_ε. In Section 3, we shall use the following bounded sets in V_ε:

$$(2.36) \qquad \begin{cases} \mathcal{B}_{\varepsilon,0} \equiv \{u = v + w : \|A_\varepsilon^{1/2} v\|^2 \le \eta_1^{-2}, \|A_\varepsilon^{1/2} w\|^2 \le \varepsilon^p \eta_3^{-2}\} \\ \mathcal{B}_{\varepsilon,1} \equiv \{u = v + w : \|A_\varepsilon^{1/2} v\|^2 \le 4\eta_1^{-2} + k_1^2 \eta_3^{-4}, \|A_\varepsilon^{1/2} w\|^2 \le k_2^2 \varepsilon^{2+r} \eta_4^{-2}\}, \\ \mathcal{B}_{\varepsilon,2} \equiv \bigcup_{t \ge 0} S_\varepsilon(\mathbf{P}_\varepsilon f, t)(\mathcal{B}_{\varepsilon,0} \cup \mathcal{B}_{\varepsilon,1}) \end{cases}$$

where $u(t) = S_\varepsilon(\mathbf{P}_\varepsilon f, t)u_0$ is the strong solution of the equation (2.13) with initial data u_0.

The inequalities (2.30) describe the norms of the data for the equation (2.13) in the space $\mathbf{L}^2(Q_3)$. We now set $p = -1$, $r = -2 + \delta$ where $\delta > 0$ is very small, and assume that $\eta_i(\varepsilon) \to 0$ as $\varepsilon \to 0^+$ for $1 \le i \le 4$. By using the mapping J_ε together with (2.7), (2.8), (2.9), we find that the hypotheses (2.30) correspond to the following conditions, for $U_0 = J_\varepsilon u_0$ and $F = J_\varepsilon f$, in the space $\mathbf{L}^2(\Omega_\varepsilon)$:

$$(2.37) \qquad \begin{cases} \|A^{1/2}\overline{M}U_0\|_{L^2(\Omega_\varepsilon)}^2 \le \varepsilon\eta_1^{-2}, \quad \|\overline{M}\mathbf{P}_3 F\|_{L^2(\Omega_\varepsilon)}^2 \le \varepsilon\eta_2^{-2}, \\ \|A^{1/2}(I - \overline{M})U_0\|_{L^2(\Omega_\varepsilon)}^2 \le \eta_3^{-2}, \quad \|(I - \overline{M})\mathbf{P}_3 F\|_{L^2(\Omega_\varepsilon)}^2 \le \varepsilon^{-1+\delta}\eta_4^{-2}, \end{cases}$$

where $\overline{M}U = \frac{1}{\varepsilon} \int_0^\varepsilon U(y_1, y_2, y_3) dy_3$.

The inequalities (2.37) imply, in particular, that

$$(2.38) \qquad \|A^{1/2}U_0\|^2_{L^2(\Omega_\epsilon)} + \|P_3F\|^2_{L^2(\Omega_\epsilon)} \le \epsilon(\eta_1^{-2} + \eta_2^{-2}) + \eta_3^{-2} + \epsilon^{-1+\delta}\eta_4^{-2}.$$

Remark 2.4. In Raugel and Sell (1990a) (section 2.9), we have recalled the classical theorem of existence of global regular solutions of the 3DNS equation (2.1) when the data are small. More precisely, in the case of our domain Ω_ϵ, it was known before that there exists a positive constant C^*, depending on ν and Q_2 (but not on ϵ), such that, if

$$(2.39) \qquad \|A^{1/2}U_0\|^2_{L^2(\Omega_\epsilon)} + \|P_3F\|^2_{L^2(\Omega_\epsilon)} \le C^*\epsilon,$$

then (2.1) has a global regular solution $U(t) \in C^0([0,\infty); D(A^{1/2}))$.

Note that the condition (2.39) also writes, on the scaled domain Q_3,

$$(2.40) \qquad \|A_\epsilon^{1/2}u_0\|^2_{L^2(Q_3)} + \|P_\epsilon f\|^2_{L^2(Q_3)} \le C^*,$$

where $U_0 = J_\epsilon u_0$, $F = J_\epsilon f$.

Since the functions $\eta_i(\epsilon) \to 0$ as $\epsilon \to 0^+$, we see that the condition (2.38) allows much bigger data than the condition (2.39). Moreover the condition (2.37) shows that the components $A^{1/2}(I - \overline{M})U_0$ and $(I - \overline{M})P_3F$ can be chosen much bigger than in (2.39).

For $0 < \epsilon \le \epsilon_0$, we define $R_1(\epsilon)$ to be the collection of $v_0 \in MV_\epsilon$ such that

$$\|A_\epsilon^{1/2}v_0\|^2 = \|v_0\|^2_{V_\epsilon} \le \eta_1^{-2},$$

and $R_2(\epsilon)$ to be the collection of $w_0 \in (I - M)V_\epsilon$ such that

$$\|A_\epsilon^{1/2}w_0\|^2 = \|w_0\|^2_{V_\epsilon} \le \epsilon^{-1}\eta_3^{-2}.$$

Set $R(\epsilon) = R_1(\epsilon) + R_2(\epsilon)$ and let $\mathcal{R}_1(\epsilon) = J_\epsilon R_1(\epsilon)$, $\mathcal{R}_2(\epsilon) = J_\epsilon R_2(\epsilon)$ and $\mathcal{R}(\epsilon) = \mathcal{R}_1(\epsilon) + \mathcal{R}_2(\epsilon) = JR(\epsilon)$ denote the corresponding sets in $(H^1(\Omega_\epsilon))^3$. The sets $R_1(\epsilon)$ and $R_2(\epsilon)$ are bounded sets in MV_ϵ and $(I - M)V_\epsilon$ with V_ϵ-radius being η_1^{-1} and $\epsilon^{-1/2}\eta_3^{-1}$, respectively. From (2.7) and (2.9), we see that $\mathcal{R}_1(\epsilon)$ and $\mathcal{R}_2(\epsilon)$ contain bounded sets in $(\overline{M}H^1(\Omega_\epsilon))^3$ and $((I - \overline{M})H^1(\Omega_\epsilon))^3$ with $H^1(\Omega_\epsilon)$-radius being $\epsilon^{1/2}\eta_1^{-1}$ and η_3^{-1}, respectively. The example 1 gives information on the size of those radii as $\epsilon \to 0^+$. The point to note in this example is that $\eta_1^{-1} = (-\log \epsilon^\alpha)^{1/4}$ and $\eta_3^{-2} = \eta_1^{-1}$. The assertion in Theorem A that $\mathcal{R}(\epsilon)$ is *large* is a heuristic formulation of the fact that $\eta_i^{-1} \to \infty$ as $\epsilon \to 0^+$, $i = 1, 3$.

Similarly we define $S(\epsilon) = S_1(\epsilon) + S_2(\epsilon)$ to be the collection of $f \in L^\infty((0,\infty),$ $L^2(Q_3))$ that satisfy

$$\|MP_\epsilon f(t)\|^2_\infty \le \eta_2^{-2} \quad \text{and} \quad \|(I - M)P_\epsilon f(t)\|^2_\infty \le \epsilon^{-2+\delta}\eta_4^{-2},$$

and set $\mathcal{S}(\epsilon) = J_\epsilon S(\epsilon) = \mathcal{S}_1(\epsilon) + \mathcal{S}_2(\epsilon)$ where $\mathcal{S}_i(\epsilon) = J_\epsilon S_i(\epsilon)$. The sets $S_1(\epsilon)$ and $S_2(\epsilon)$ are bounded sets in $ML^2(Q_3)$ and $(I - M)L^2(Q_3)$ with $L^2(Q_3)$-radius being η_2^{-1} and $\epsilon^{-1+\frac{\delta}{2}}\eta_4^{-1}$, respectively. And the sets $\mathcal{S}_1(\epsilon)$ and $\mathcal{S}_2(\epsilon)$ contain bounded sets in $\overline{M}(L^2(\Omega_\epsilon))^3$ and $(I - \overline{M})(L^2(\Omega_\epsilon))^3$ with $L^2(\Omega_\epsilon)$-radius being $\epsilon^{1/2}\eta_2^{-1}$ and $\epsilon^{-1/2+\delta/2}\eta_4^{-1}$, respectively. Once again, the assertion that $\mathcal{S}(\epsilon)$ is large is a heuristic formulation of the fact that $\eta_i^{-1} \to \infty$ as $\epsilon \to 0^+$, $i = 2, 4$. The example 1 shows that one can choose $\eta_2^{-1} = (\log(-\log \epsilon^\alpha)^\beta)^{1/4}$ and $\eta_4^{-1} = (-\log \epsilon)$.

2.6 Statement of the regularity results in the purely periodic case. In the purely periodic case, i.e., where (2.2) holds, we have global existence of strong solutions for bigger data than those described in Theorem 1. Let us now give these results, proved in Raugel and Sell (1990b). We shall say that the bounded monotone functions $\eta_i(\varepsilon)$ defined for $0 < \varepsilon \leq 1$, $i = 1, 2, 3, 4$ and the *negative* constants p, q_1, q_2, r satisfy *Hypothesis 1*, provided

$$
\begin{cases}
(1) & r > -2, p > -\frac{29}{24}, \quad q_1 > -\frac{5}{12}, q_2 > -\frac{5}{12} \\
(2) & \text{there exist positive constants } \alpha_i, \ \beta_i, \ i = 1, 2, 3, 4 \\
& \text{such that } \beta_i \leq \eta_i^{-1}(\varepsilon) < (-\log \varepsilon)^{\alpha_i}.
\end{cases}
$$

Since we are interested in big data, we assume that p, q_1, q_2, r are negative numbers. Let us give an example where these conditions hold.

Example 2. We set

$$
\begin{cases}
p = -\frac{29}{24} + \delta_1, \quad q_1 = -\frac{5}{12} + \delta_2, \quad q_2 = -\frac{5}{12} + \delta_3 \\
r = -2 + \delta_4,
\end{cases}
$$

where δ_i are arbitrarily small positive numbers, $\eta_i^{-1}(\varepsilon) = (-\log \varepsilon)^{\alpha_i}$, $\alpha_i > 0$, $i = 1, 2, 3, 4$. Then the *Hypothesis 1* holds.

In Raugel and Sell (1990b), we have shown the following result.

THEOREM 2. *Assume that the Navier-Stokes equations (2.1) are supplemented with the periodic boundary conditions (2.2). Let η_i, $i = 1, 2, 3, 4$, p, q_1, q_2, r satisfy Hypothesis 1. Then there exist a positive number ε_0^*, a positive constant C^*, which is independent of ε and t, and, for all ε, $0 < \varepsilon \leq \varepsilon_0^*$, a time $\widehat{T}_1^* = \widehat{T}_1^*(\varepsilon) > 0$, such that, whenever $u_0 \in D(A_\varepsilon^{1/2})$, $f \in L^\infty((0, \infty), L^2(Q_3))$ satisfy*

$$
(2.41) \qquad
\begin{aligned}
& \|A_\varepsilon^{1/2} v_0\|^2 \leq \varepsilon^{q_1} \eta_1^{-2}, \|MP_\varepsilon f\|_\infty^2 \leq \varepsilon^{q_2} \eta_2^{-2}, \\
& \|A_\varepsilon^{1/2} w_0\|^2 \leq \varepsilon^p \eta_3^{-2}, \|(I - M)P_\varepsilon f\|_\infty^2 \leq \varepsilon^r \eta_4^{-2},
\end{aligned}
$$

then (2.13) has a (unique) strong solution u that belongs to $C^0([0, \infty), V_\varepsilon) \cap L^\infty([0, \infty), V_\varepsilon)$, i.e., one has

$$
(2.42) \qquad \|A_\varepsilon^{1/2} u(t)\|^2 \leq K_1^{*2}, \quad t \geq 0,
$$

where K_1^* is independent of t. Moreover the components of $u = v + w$ satisfy

$$
(2.43) \qquad \|A_\varepsilon^{1/2} v(t)\|^2 \leq C^*(\varepsilon^{q_2} \eta_2^{-2} + 1 + \varepsilon^{2+r} \eta_4^{-2} + \varepsilon^{4+2r} \eta_4^{-4})^3, \quad t \geq \widehat{T}_1^*
$$

and

$$
(2.44) \qquad \|A_\varepsilon^{1/2} w(t)\|^2 \leq k_2^2 \varepsilon^{2+r} \eta_4^{-2}, \quad t \geq \widehat{T}_1^*
$$

where k_2 depends only on ν and Q_2.

The remark 2.1 is still true, in this case.

Remark 2.5. As for Theorem 1, the proof of Theorem 2 is made in two steps (see Raugel and Sell 1990b)). The first step consists in showing that if u_0 and f satisfy (2.41), there is a positive time $T_1^* = T_1^*(\varepsilon)$ such that $u(t) \in D(A_\varepsilon^{1/2})$ for $0 \leq t \leq T_1^*$ and

$$(2.45) \qquad \begin{cases} \|A_\varepsilon^{1/2} v(T_1^*)\|^2 \leq 4\varepsilon^{q_1}\eta_1^{-2} + k_1^2\varepsilon^{2+2p}\eta_3^{-4}, \\ \|A_\varepsilon^{1/2} w(T_1^*)\|^2 \leq k_2^2\varepsilon^{2+r}\eta_4^{-2}, \end{cases}$$

where k_1 is a positive constant depending only on ν and Q_2.

The second step consists in showing that if the data u_0 and f satisfy (2.45) and (2.41) respectively, then (2.13) has a (unique) strong solution u in $C^0([0,\infty), V_\varepsilon) \cap L^\infty((0,\infty), V_\varepsilon)$ and the estimates (2.43) and (2.44) hold. The difference in the proofs of Theorems 1 and 2 comes from the fact that, in the case of the periodic boundary conditions (2.2), we use the following decomposition of $u(t)$

$$(2.46) \qquad u(t) = Mu(t) + (I - M)u(t) \equiv v(t) + w(t),$$

$$(2.47) \qquad v(t) = \tilde{v}(t) + v^*(t) = (v_1(t), v_2(t), 0) + (0, 0, v_3(t))$$
$$\text{where } v(t) = (v_1(t), v_2(t), v_3(t)),$$

and we use the conservation of enstrophy property

$$(2.48) \qquad b_\varepsilon(\tilde{v}, \tilde{v}, A_\varepsilon \tilde{v}) = 0.$$

In Section 3, we shall use the following bounded sets in $V_\varepsilon = V_\varepsilon^1 = D(A_\varepsilon^{1/2})$:

$$(2.49) \qquad \begin{cases} \mathcal{B}_{\varepsilon,0}^* \equiv \left\{ u = v + w : \|A_\varepsilon^{1/2} v\|^2 \leq \varepsilon^{\tilde{q}_1}\tilde{\eta}_1^{-2}, \right. \\ \qquad\qquad \left. \|A_\varepsilon^{1/2} w\|^2 \leq \varepsilon^p \eta_3^{-2} \right\} \\ \mathcal{B}_{\varepsilon,1}^* \equiv \left\{ u = v + w : \|A_\varepsilon^{1/2} v\|^2 \leq 4\varepsilon^{\tilde{q}_1}\tilde{\eta}_1^{-2} + k_1^2\varepsilon^{2+2p}\eta_3^{-4}, \right. \\ \qquad\qquad \left. \|A_\varepsilon^{1/2} w\|^2 \leq k_2^2\varepsilon^{2+r}\eta_4^{-2} \right\} \\ \mathcal{B}_{\varepsilon,2}^* = \bigcup_{t \geq 0} S_\varepsilon(\mathbf{P}_\varepsilon f, t)(\mathcal{B}_{\varepsilon,0}^* \cup \mathcal{B}_{\varepsilon,1}^*). \end{cases}$$

where $u(t) = S_\varepsilon(\mathbf{P}_\varepsilon f, t)u_0$ is the strong solution of the equation (2.13) with initial data u_0, and where

$$\tilde{q}_1 = \min(q_1, q_2), \quad \tilde{\eta}_1^{-2} = \max(8\nu^{-2}\lambda_1^{-1}\eta_2^{-2}, \eta_1^{-2}).$$

Let us go back to the example 2. By using the mapping J_ε together with (2.7), (2.8), (2.9), we find that the hypotheses (2.41), in this case, correspond to the following conditions, for $U_0 = J_\varepsilon u_0$ and $F = J_\varepsilon f$, in the space $L^2(\Omega_\varepsilon)^3$:

$$(2.50) \qquad \begin{cases} \|A^{1/2}\overline{M}U_0\|_{L^2(\Omega_\varepsilon)}^2 \leq \varepsilon^{\frac{7}{12}+\delta_2}\eta_1^{-2}, \|\overline{M}\mathbf{P}_3 F\|_{L^2(\Omega_\varepsilon)}^2 \leq \varepsilon^{\frac{7}{12}+\delta_3}\eta_2^{-2}, \\ \|A^{1/2}(I - \overline{M})U_0\|_{L^2(\Omega_\varepsilon)}^2 \leq \varepsilon^{-\frac{5}{24}+\delta_1}\eta_3^{-2}, \|(I - \overline{M})\mathbf{P}_3 F\|_{L^2(\overline{M}_\varepsilon)}^2 \\ \qquad\qquad\qquad\qquad\qquad\qquad \leq \varepsilon^{-1+\delta_4}\eta_4^{-2}. \end{cases}$$

The inequalities (2.50) imply, in particular, that

$$
\|A^{1/2}U_0\|^2_{L^2(\Omega_\varepsilon)} + \|\mathbf{P}_3 F\|^2_{L^2(\Omega_\varepsilon)} \leq \varepsilon^{\frac{7}{12}+\delta_2}\eta_1^{-2} + \varepsilon^{-\frac{5}{24}+\delta_1}\eta_3^{-2}
$$
$$
\qquad\qquad + \varepsilon^{\frac{7}{12}+\delta_3}\eta_2^{-2} + \varepsilon^{-1+\delta_4}\eta_4^{-2}.
$$

(2.51)

The condition (2.50) shows that the components $A^{1/2}(I - \overline{M})U_0$ and $(I - \overline{M})\mathbf{P}_3 F$ can be chosen much bigger than in (2.39).

For $0 < \varepsilon \leq \varepsilon_0^*$, we define $R_1^*(\varepsilon)$ to be the collection of $v_0 \in MV_\varepsilon$ such that

$$
\|A_\varepsilon^{1/2} v_0\|^2 \leq \varepsilon^{-\frac{5}{12}+\delta_2}\eta_1^{-2},
$$

and $R_2^*(\varepsilon)$ to be the collection of $w_0 \in (I - M)V_\varepsilon$ such that

$$
\|A_\varepsilon^{1/2} w_0\|^2 \leq \varepsilon^{-\frac{23}{24}+\delta_1}\eta_3^{-2}.
$$

Set $R^*(\varepsilon) = R_1^*(\varepsilon) + R_2^*(\varepsilon)$ and let $\mathcal{R}_1^*(\varepsilon) = J_\varepsilon R_1^*(\varepsilon)$, $\mathcal{R}_2^*(\varepsilon) = J_\varepsilon R_2^*(\varepsilon)$ and $\mathcal{R}^*(\varepsilon) = \mathcal{R}_1^*(\varepsilon) + \mathcal{R}_2^*(\varepsilon) = J_\varepsilon R^*(\varepsilon)$ denote the corresponding sets in $(H^1(\Omega_\varepsilon))^3$. The sets $\mathcal{R}_1^*(\varepsilon)$ and $\mathcal{R}_2^*(\varepsilon)$ contain bounded sets in $\overline{M}(H^1(\Omega_\varepsilon))^3$ and $(I - \overline{M})(H^1(\Omega_\varepsilon))^3$ with $H^1(\Omega_\varepsilon)$-radius being $\varepsilon^{\frac{7}{24}+\frac{1}{2}\delta_2}\eta_1^{-1}$ and $\varepsilon^{-\frac{5}{48}+\frac{1}{2}\delta_1}\eta_3^{-1}$, respectively. We recall that $\eta_i^{-1} = (-\log\varepsilon)^{\alpha_i}$, with $\alpha_i > 0$, $i = 1, 2, 3, 4$.

Similarly we define $S^*(\varepsilon) = S_1^*(\varepsilon) + S_2^*(\varepsilon)$ to be the collection of $f \in L^\infty((0,\infty), \mathbf{L}^2(Q_3))$ that satisfy

$$
\|M\mathbf{P}_\varepsilon f(t)\|^2_\infty \leq \varepsilon^{-\frac{5}{12}+\delta_3}\eta_2^{-2}
$$

and

$$
\|(I - M)\mathbf{P}_\varepsilon f(t)\|^2_\infty \leq \varepsilon^{-2+\delta_4}\eta_4^{-2},
$$

and set $S^*(\varepsilon) = J_\varepsilon S^*(\varepsilon) = S_1^*(\varepsilon) + S_2^*(\varepsilon)$ where $S_i^*(\varepsilon) = J_\varepsilon S_i^*(\varepsilon)$. The sets $S_1^*(\varepsilon)$ and $S_2^*(\varepsilon)$ contain bounded sets in $\overline{M}(L^2(\Omega_\varepsilon))^3$ and $(I-\overline{M})(L^2(\Omega_\varepsilon))^3$ with $L^2(\Omega_\varepsilon)$-radius being $\varepsilon^{\frac{7}{24}+\frac{1}{2}\delta_3}\eta_2^{-1}$ and $\varepsilon^{-1+\frac{1}{2}\delta_4}\eta_4^{-1}$, respectively.

If we allow the third components of $A_\varepsilon^{1/2}v_0$ and $M\mathbf{P}_\varepsilon f$ to be *small*, we obtain a better result of existence of strong solutions in the case of the periodic boundary conditions (2.2). Let us now state this result. We shall say that the bounded monotone functions $\eta_i(\varepsilon)$ defined for $0 < \varepsilon \leq 1$, $i = 1, 2, 3, 4$, the *negative* constants p, q_1, q_2, r and the positive numbers m_1, m_2 satisfy *Hypothesis 2*, provided

(1) $0 \leq m_1 < 2, 0 \leq m_2$,

(2) $-2 < r < 0, -\frac{5}{4} < q_1 < 0, -\frac{5}{4} < q_2 < 0, -\frac{5}{4} < p < 0$,

(3) $p > \frac{1}{2}m_1 - 2$

(4) $2q_i + \inf(m_1, m_2) + \frac{5}{4} > 0$, $i = 1, 2$

(5) $4q_1 + 2m_1 + \inf(m_1, m_2) + \frac{5}{4} > 0$,

(6) $8 + 8p + 2m_1 + \inf(m_1, m_2) + \frac{5}{4} > 0$, and

(7) there exist positive constants $\alpha_i, \beta_i, i = 1, 2, 3, 4$ such that

$$\beta_i \leq \eta_i^{-1}(\varepsilon) \leq (-\log \varepsilon)^{\alpha_i}$$

Again, since we are interested in big data, we assume that p, q_1, q_2, r are negative numbers.

Let us give an example where *Hypothesis 2* holds.

Example 3. If we set

$$m_1 = m_2 = 1, \qquad p = -\frac{5}{4} + \delta_1,$$

$$q_1 = -\frac{17}{16} + \delta_2, \quad q_2 = -\frac{9}{8} + \delta_3, \quad r = -2 + \delta_4,$$

where δ_i are arbitrarily small positive numbers and $\eta_i^{-1}(\varepsilon) = (-\log \varepsilon)^{\alpha_i}$, $\alpha_i > 0$, $i = 1, 2, 3, 4$ then *Hypothesis 2* holds. This example shows that if m_1, m_2 are big enough, we can choose $q_1 < -1$ and $q_2 < -1$.

In Raugel and Sell (1990b), we prove the following result:

We recall that, if $v \in ML^2(Q_3)$, we denote by v^* the vector $(0, 0, v_3)$ where $v = (v_1, v_2, v_3)$.

THEOREM 3. *Assume that the Navier-Stokes equations (2.1) are supplemented with the periodic boundary conditions (2.2). Let η_i, $i = 1, 2, 3, 4$, p, q_1, q_2, r, m_1, m_2 satisfy Hypothesis 2 and let K_1, K_2 be two positive constants. Then there exist a positive number $\widetilde{\varepsilon}_0$, a positive constant $\widetilde{C}$, which is independent of ε and t, and, for all ε, $0 < \varepsilon \leq \widetilde{\varepsilon}_0$, a time $\widetilde{T}_1 = \widetilde{T}_1(\varepsilon) > 0$, such that, whenever $u_0 \in D(A_\varepsilon^{1/2})$, $f \in L^\infty((0, \infty); L^2(Q_3))$ satisfy*

$$(2.52) \qquad \begin{cases} \|A_\varepsilon^{1/2} v_0\|^2 \leq \varepsilon^{q_1} \eta_1^{-2}, \|MP_\varepsilon f\|_\infty^2 \leq \varepsilon^{q_2} \eta_2^{-2}, \\ \|A_\varepsilon^{1/2} w_0\|^2 \leq \varepsilon^p \eta_3^{-2}, \|(I - M)P_\varepsilon f\|_\infty^2 \leq \varepsilon^r \eta_4^{-2}, \end{cases}$$

and

$$(2.53) \qquad \|v_0^*\|^2 \leq K_1^2 \varepsilon^{m_1}, \quad \|(MP_\varepsilon f)^*\|_\infty^2 \leq K_2^2 \varepsilon^{m_2},$$

then (2.13) has a (unique) strong solution u that belongs to $C^0([0, \infty), V_\varepsilon) \cap L^\infty((0, \infty), V_\varepsilon)$, i. e., one has

$$(2.54) \qquad \|A_\varepsilon^{1/2} u(t)\|^2 \leq \widetilde{K}_1^2, \quad t \geq 0,$$

where $\widetilde{K}_1$ is independent of t. Moreover the components of $u = v + w$ satisfy

$$(2.55) \qquad \|A_\varepsilon^{1/2} v(t)\|^2 \leq \widetilde{C}(1 + \varepsilon^{q_2} \eta_2^{-2} + \varepsilon^{2q_2} \eta^{-4}(\varepsilon^{m_2} + \varepsilon^2)), \quad t \geq \widetilde{T}_1$$

and

$$(2.56) \qquad \|A_\varepsilon^{1/2} w(t)\|^2 \leq k_2^2 \varepsilon^{2+r}, \quad t \geq \widetilde{T}_1,$$

where k_2 depends only on ν and Q_2.

Let us go back to the example 3. By using the mapping J_ε together with (2.7), (2.8), (2.9), we find that the hypotheses (2.52) and (2.53), in this case, correspond to the following conditions, for $U_0 = J_\varepsilon u_0$ and $F = J_\varepsilon f$, in the space $L^2(\Omega_\varepsilon)$:

(2.57)
$$\begin{cases} \|A^{1/2}\overline{M}U_0\|^2_{L^2(\Omega_\varepsilon)} \leq \varepsilon^{-\frac{1}{16}+\delta_2}\eta_1^{-2}, \|\overline{M}P_3F\|^2_{L^2(\Omega_\varepsilon)} \leq \varepsilon^{-\frac{1}{8}+\delta_3}\eta_2^{-2}, \\ \|A^{1/2}(I-\overline{M})U_0\|^2_{L^2(\Omega_\varepsilon)} \leq \varepsilon^{-\frac{1}{4}+\delta_1}\eta_3^{-2}, \|(I-\overline{M})P_3F\|^2_{L^2(\Omega_\varepsilon)} \leq \varepsilon^{-1+\delta_4}\eta_4^{-2}, \end{cases}$$

(2.58)
$$\|(\overline{M}U_0)^*\|^2_{L^2(\Omega_\varepsilon)} \leq K_1^2\varepsilon^2, \|(\overline{M}P_3F)^*\|^2_\infty \leq K_2^2\varepsilon^2.$$

The inequalities (2.57) imply, in particular, that

(2.59)
$$\|A^{1/2}U_0\|^2_{L^2(\Omega_\varepsilon)} + \|P_3F\|^2_{L^2(\Omega_\varepsilon)} \leq \varepsilon^{-\frac{1}{16}+\delta_2}\eta_1^{-2}$$
$$+ \varepsilon^{-\frac{1}{4}+\delta_1}\eta_3^{-2} + \varepsilon^{-1/8+\delta_3}\eta_2^{-2} + \varepsilon^{-1+\delta_4}\eta_4^{-2}.$$

The conditions (2.59) show that, under the hypothesis (2.58), the data $A^{1/2}U_0$ and P_3F can be chosen much bigger than in (2.39).

For $0 < \varepsilon \leq \tilde{\varepsilon}_0$, we define $\tilde{R}_1(\varepsilon)$ to be the collection of $v_0 \in MV_\varepsilon$ such that

$$\|A_\varepsilon^{1/2}v_0\|^2 \leq \varepsilon^{-\frac{17}{16}+\delta_2}\eta_1^{-2}, \quad \|v_0^*\|^2 \leq K_1^2\varepsilon$$

and $\tilde{R}_2(\varepsilon)$ to be the collection of $w_0 \in (I-M)V_\varepsilon$ such that

$$\|A_\varepsilon^{1/2}w_0\|^2 \leq \varepsilon^{-\frac{5}{4}+\delta_1}\eta_3^{-2}.$$

Set $\tilde{R}(\varepsilon) = \tilde{R}_1(\varepsilon) + \tilde{R}_2(\varepsilon)$ and let $\tilde{\mathcal{R}}_1(\varepsilon) = J_\varepsilon\tilde{R}_1(\varepsilon)$, $\tilde{\mathcal{R}}_2(\varepsilon) = J_\varepsilon\tilde{R}_2(\varepsilon)$ and $\tilde{\mathcal{R}}_\varepsilon = \tilde{\mathcal{R}}_1(\varepsilon) + \tilde{\mathcal{R}}_2(\varepsilon) = J_\varepsilon\tilde{\mathcal{R}}(\varepsilon)$ denote the corresponding sets in $(H^1(\Omega_\varepsilon))^3$. Note that the sets $\tilde{\mathcal{R}}_1(\varepsilon)$ and $\tilde{\mathcal{R}}_2(\varepsilon)$ contain bounded sets in $\overline{M}(H^1(\Omega_\varepsilon))^3$ and $(I-\overline{M})(H^1(\Omega_\varepsilon))^3$ with $H^1(\Omega_\varepsilon)$-radius being $\varepsilon^{-\frac{1}{32}+\frac{\delta_2}{2}}\eta_1^{-2}$ and $\varepsilon^{-\frac{1}{8}+\frac{\delta_1}{2}}\eta_3^{-1}$, respectively. These radii tend to infinity as ε goes to 0.

Similarly, we define $\tilde{S}(\varepsilon) = \tilde{S}_1(\varepsilon) + \tilde{S}_2(\varepsilon)$ to be the collection of $f \in L^\infty((0,\infty), L^2(Q_3))$ that satisfy
$$\|MP_\varepsilon f(t)\|^2_\infty \leq \varepsilon^{-\frac{9}{8}+\delta_3}\eta_2^{-2}, \|(MP_\varepsilon f)^*\|^2_\infty \leq K_2^2\varepsilon$$

and
$$\|(I-M)P_\varepsilon f(t)\|^2_\infty \leq \varepsilon^{-2+\delta_4}\eta_4^{-2},$$

and set $\tilde{S}(\varepsilon) = J_\varepsilon\tilde{S}(\varepsilon) = \tilde{S}_1(\varepsilon) + \tilde{S}_2(\varepsilon)$ where $\tilde{S}_i(\varepsilon) = J_\varepsilon\tilde{S}_i(\varepsilon)$. The sets $\tilde{S}_1(\varepsilon)$ and $\tilde{S}_2(\varepsilon)$ contain bounded sets in $\overline{M}(L^2(\Omega_\varepsilon))^3$ and $(I-\overline{M})(L^2(\Omega_\varepsilon))^3$ with $L^2(\Omega_\varepsilon)$-radius being $\varepsilon^{-\frac{1}{16}+\frac{\delta_3}{2}}\eta_2^{-1}$ and $\varepsilon^{-1/2+\frac{\delta_4}{2}}\eta_4^{-1}$, respectively. These radii tend to infinity as ε goes to 0.

3. Existence of a global attractor.

3.1 Local attractors. *Here we assume that the forcing term f (or F) is independent of the time t.* Recall that we denote by $u(t) = S_\varepsilon(\mathbf{P}_\varepsilon f, t)u_0$ the strong solution of the equation (2.13) with initial data u_0. Sometimes, if no confusion is possible, we denote $S_\varepsilon(\mathbf{P}_\varepsilon f, t)$ simply by $S_\varepsilon(t)$. If f is independent of t and satisfies the hypotheses of Theorem 1, $S_\varepsilon(\mathbf{P}_\varepsilon f, t)$ is a C^0-semigroup, for $t \geq 0$, from $\mathcal{B}_{\varepsilon,2}$ into $\mathcal{B}_{\varepsilon,2}$. In the case of the boundary conditions (2.2), if f is independent of t and satisfies the hypotheses of Theorem 2, $S_\varepsilon(\mathbf{P}_\varepsilon f, t)$ is a C^0-semigroup for $t \geq 0$, from $\mathcal{B}_{\varepsilon,2}^*$ into $\mathcal{B}_{\varepsilon,2}^*$.

Assume that f satisfies the hypotheses of Theorem 1. Let $\mathcal{A}_\varepsilon = \omega(\mathcal{B}_{\varepsilon,2})$ be the ω-limit set of $\mathcal{B}_{\varepsilon,2}$ in V_ε, i.e.,

$$\mathcal{A}_\varepsilon \equiv \bigcap_{\tau \geq 0} \mathrm{Closure}_{V_\varepsilon} \left(\bigcup_{t \geq \tau} S_\varepsilon(\mathbf{P}_\varepsilon f, t)\mathcal{B}_{\varepsilon,2} \right).$$

It follows from Theorem 2 of [Raugel and Sell, 1990a] that for $\tau \geq \widehat{T}_1 + 1$, the set

$$\bigcup_{t \geq \tau} S_\varepsilon(\mathbf{P}_\varepsilon f, t)\mathcal{B}_{\varepsilon,2}$$

lies in a bounded set in V_ε^2, and thus in a non-empty compact set in V_ε. Consequently $\mathcal{A}_\varepsilon$ is a nonempty compact invariant set in V_ε, attracting $\mathcal{B}_{\varepsilon,2}$ in V_ε. Since

$$S_\varepsilon(\mathbf{P}_\varepsilon f, t)\mathcal{B}_{\varepsilon,2} \subset \mathcal{B}_{\varepsilon,2},$$

$\mathcal{A}_\varepsilon$ is a local attractor for the strong solutions of (2.13) in V_ε and the basin of attraction $\mathcal{B}(\mathcal{A}_\varepsilon)$ satisfies:

$$\mathcal{B}_{\varepsilon,2} \subset \mathcal{B}(\mathcal{A}_\varepsilon).$$

Remark that $\mathcal{A}_\varepsilon$ is the maximal attractor of the restriction of $S_\varepsilon(\mathbf{P}_\varepsilon f, t)$ to $\mathcal{B}_{\varepsilon,2}$.

We thus obtain the following theorem.

THEOREM 3.1. *Let $\eta_i, i = 1, 2, 3, 4, r < 0$ and $p < 0$ satisfy Hypothesis $H(a,b)$, where a and b are sufficiently large. Assume that f is time-independent and satisfies*

$$(3.1) \qquad \|M\mathbf{P}_\varepsilon f\|_\infty^2 \leq \eta_2^{-2}, \quad \|(I - M)\mathbf{P}_\varepsilon f\|_\infty^2 \leq \varepsilon^r \eta_4^{-2}.$$

Let $\varepsilon_0 > 0$ be given by Theorem 1. Then, for $0 < \varepsilon \leq \varepsilon_0$, the semigroup $S_\varepsilon(\mathbf{P}_\varepsilon f, t)$ generated by the strong solutions of the dilated 3D Navier-Stokes evolutionary equation (2.13) has a unique, maximal compact (local) attractor $\mathcal{A}_\varepsilon$, included in $\mathcal{B}_{\varepsilon,2}$, which attracts $\mathcal{B}_{\varepsilon,2}$ in the space V_ε. Moreover,

$$(3.2) \qquad \mathcal{A}_\varepsilon \subset \left\{ u = v + w : \|A_\varepsilon^{1/2} v(t)\|^2 \leq \Gamma(\eta_2^{-2}), \|A_\varepsilon^{1/2} w(t)\|^2 \leq k_2^2 \varepsilon^{2+r} \eta_4^{-2} \right\}$$

Moreover, $\mathcal{A}_\varepsilon$ is bounded in V_ε^2 and attracts the bounded set $\mathcal{B}_{\varepsilon,2} \cap V_\varepsilon^2$ in the space V_ε^2.

The property (3.2) is a direct consequence of Theorem 1; the last assertion is proved in Raugel and Sell (1990a). In the case of the periodic boundary conditions (2.2), the same type of argument leads to the following result.

THEOREM 3.2. *Assume that the Navier-Stokes equations (2.1) are supplemented with the periodic boundary conditions (2.2) Let η_i, $i = 1, 2, 3, 4$, $p < -1, q_1 < 0, q_2 < 0, r < 0$ satisfy Hypothesis 1. Assume that f is time-independent and satisfies*

$$(3.3) \qquad \|MP_\varepsilon f\|_\infty^2 \leq \varepsilon^{q_2} \eta_2^{-2}, \quad \|(I - M)P_\varepsilon f\|_\infty^2 \leq \varepsilon^r \eta_4^{-2}.$$

Let $\varepsilon_0^* > 0$ be given in Theorem 2. Then, for $0 < \varepsilon \leq \varepsilon_0^*$, the semigroup $S_\varepsilon(P_\varepsilon f, t)$ generated by the strong solutions of the dilated 3D Navier-Stokes evolutionary equation (2.13) has a unique, maximal compact (local) attractor $\mathcal{A}_\varepsilon^*$, included in $\mathcal{B}_{\varepsilon,2}^*$, which attracts $\mathcal{B}_{\varepsilon,2}^*$ in the space V_ε. Moreover,

$$(3.4)$$
$$\mathcal{A}_\varepsilon^* \subset \left\{ u = v + w : \|A_\varepsilon^{1/2} v(t)\|^2 \leq C^*(1 + G_2(\varepsilon))^3, \|A_\varepsilon^{1/2} w(t)\|^2 \leq k_2^2 \varepsilon^{2+r} \eta_4^{-2} \right\},$$

where

$$(3.5) \qquad G_2(\varepsilon) = \varepsilon^{q_2} \eta_2^{-2} + \varepsilon^{2+r} \eta_r^{-2} + \varepsilon^{4+2r} \eta_4^{-4}$$

Moreover, $\mathcal{A}_\varepsilon^*$ is bounded in V_ε^2 and attracts the bounded set $\mathcal{B}_{\varepsilon,2}^* \cap V_\varepsilon^2$ in the space V_ε^2.

3.2 Global attractors.

In the next theorems, we show that, under an additional condition on η_i, $i = 1, 2, 3, 4$ (see (3.6) below), the local attractor $\mathcal{A}_\varepsilon$ is the global attractor for the weak (Leray) solutions of the dilated 3D Navier-Stokes evolutionary equation (2.13) (see (1.5), (1.6) and (1.7) in the introduction). Note that the example 1, given prior Theorem 1, satisfies the condition (3.6).

COROLLARY 3.1. *Let the hypotheses of Theorem 3.1 be satisfied. Assume, in addition, that, for every $\lambda > 0$, there exists a positive number ε_1 such that, for $0 < \varepsilon \leq \min(\varepsilon_0, \varepsilon_1)$,*

$$(3.6) \qquad \eta_2^{-2} + \varepsilon^{2+r} \eta_4^{-2} \leq \lambda \min(\eta_1^{-2}, \varepsilon^p \eta_3^{-2}),$$

where $\lambda < \nu^2 (2 \max(C_3^2, \lambda_1^{-1}))^{-1}$, where $\lambda_1 > 0$ is the smallest eigenvalue of A_ε.

Then, for $0 < \varepsilon \leq \min(\varepsilon_0, \varepsilon_1)$, for any $\rho > 0$, there exists a time $T_\varepsilon = T_\varepsilon(\rho)$ and, for every weak (Leray) solution $u(t)$ of (2.13) with $\|u(0)\| \leq \rho$, there is a positive time t_0, $0 < t_0 \leq T_\varepsilon(\rho)$ such that

$$u(t_0) \in \mathcal{B}_{\varepsilon,2}.$$

In particular, $u(t)$ is a strong solution of (2.13) for $t \geq t_0$, and the (local) attractor $\mathcal{A}_\varepsilon$ given in Theorem 3.1 is the global attractor for the weak (Leray) solutions of (2.13), provided $0 < \varepsilon \leq \min(\varepsilon_0, \varepsilon_1)$.

Remark 3.2. Corollary 3.1 implies that the weak global attractor $\mathcal{X}$ coincides with $\mathcal{A}_\varepsilon$. Thus we have a positive answer to the question (1.11): $\mathcal{X}$ is contained in V_ε.

Proof of Corollary 3.1. For any weak (Leray) solution $u(t)$ of (2.13), we deduce from (1.7) that, for almost every $t > 0$,

$$\nu \int_0^t \|A_\varepsilon^{1/2} u\|^2 ds \leq \|u(0)\|^2 + 2t \left(\|A_\varepsilon^{-1/2} M \mathbf{P}_\varepsilon f\|^2 + \|A_\varepsilon^{-1/2}(I - M)\mathbf{P}_\varepsilon f\|^2 \right),$$

which implies, thanks to the inequality (2.22)(ii) and the condition (3.6),

$$\frac{1}{t} \int_0^t \|A_\varepsilon^{1/2} u(s)\|^2 ds \leq \frac{\nu^{-1}}{t} \|u(0)\|^2 + 2\nu^{-2} \left(\lambda_1^{-1} \|M \mathbf{P}_\varepsilon f\|^2 + C_3^2 \varepsilon^2 \|(I - M)\mathbf{P}_\varepsilon f\|^2 \right)$$

$$\leq \frac{\nu^{-1}}{t} \|u(0)\|^2 + 2\nu^{-2} \left(\lambda_1^{-1} \eta_2^{-2} + C_3^2 \varepsilon^{2+r} \eta_4^{-2} \right)$$

$$\leq \frac{\nu^{-1}}{t} \|u(0)\|^2 + k \min \left(\eta_1^{-2}, \varepsilon^p \eta_3^{-2} \right),$$

for $0 < \varepsilon \leq \min(\varepsilon_0, \varepsilon_1)$, where $0 < k < 1$. Since $\|u(0)\| \leq \rho$, there is a positive time t_0, $0 < t_0 \leq T(\rho)$, where

$$(3.9) \qquad\qquad T(\rho) = \frac{2\nu^{-1}\rho^2}{(1-k)\min(\eta_1^{-2}, \varepsilon^p \eta_3^{-2})},$$

such that,

$$(3.10) \qquad\qquad \|A_\varepsilon^{1/2} u(t_0)\|^2 \leq \min(\eta_1^{-2}, \varepsilon^p \eta_3^{-2}).$$

For this time t_0, it follows from (3.10), since M is also an orthogonal projector in V_ε^1,

$$(3.11) \qquad \begin{cases} \|A_\varepsilon^{1/2} v(t_0)\|^2 \leq \|A_\varepsilon^{1/2} u(t_0)\|^2 \leq \eta_1^{-2}, \\ \|A_\varepsilon^{1/2} w(t_0)\|^2 \leq \|A_\varepsilon^{1/2} u(t_0)\|^2 \leq \varepsilon^p \eta_3^{-2}, \end{cases}$$

and the corollary is proved.

In the case of the *periodic boundary conditions (2.2)*, the same argument leads to the following result.

COROLLARY 3.2. *Let the hypotheses of Theorem 3.2 be satisfied. Then, there exists a positive number ε_1^* such that, for $0 < \varepsilon \leq \min(\varepsilon_0^*, \varepsilon_1^*)$, for any $\rho > 0$, there exists a time $T_\varepsilon^* = T_\varepsilon^*(\rho)$ and, for every weak (Leray) solution $u(t)$ of (2.13) with $\|u(0)\| \leq \rho$, there is a positive time t_0, $0 < t_0 \leq T_\varepsilon^*(\rho)$ such that*

$$u(t_0) \in \mathcal{B}_{\varepsilon,2}^*.$$

In particular, $u(t)$ is a strong solution of (2.13) for $t \geq t_0$, and the (local) attractor $\mathcal{A}_\varepsilon^*$ given in Theorem 3.2 is the global attractor for the weak (Leray) solutions of (2.13), provided $0 < \varepsilon \leq \min(\varepsilon_0^*, \varepsilon_1^*)$.

3.3 A particular case. Let us now consider the reduced 3D Navier-Stokes evolutionary equation (2.24) and let us denote by $S_0(g, t)\bar{v}_0 = \bar{v}(t)$ the strong solution of (2.24) with initial data $\bar{v}_0$ in MV_ε^1, where $g = M\mathbf{P}_\varepsilon f$ (We will simply denote $S_0(g, t)$ by $S_0(t)$, if no confusion is possible). In [Raugel and Sell, 1990a], we have shown that $S_0(t)$ is a C^0-semigroup from MV_ε^1 into MV_ε^1. We have also shown that $S_0(t)$ has a global attractor $\mathcal{A}_0 \equiv \mathcal{A}_0(g)$ in MV_ε^1 (see [Section 2.13 in Raugel and Sell, 1990a]). Moreover, we have proved the following result.

PROPOSITION 3.3. *Assume that the hypotheses of Theorem 3.1 hold, then the global attractor $\mathcal{A}_0 = \mathcal{A}_0(g)$ of $S_0(g, t)$ in MV_ε^1 satisfies*

$$(3.12) \qquad \mathcal{A}_0(g) \subset \{u = \bar{v} + w : v \in MV_\varepsilon^1, \|A_\varepsilon^{1/2}\bar{v}\|^2 \le \Gamma(\eta_2^{-2}), w = 0\}$$

If, in addition, one has

$$(I - M)\mathbf{P}_\varepsilon f = 0,$$

then the attractors $\mathcal{A}_\varepsilon$ and $\mathcal{A}_0(g)$ coincide for $0 < \varepsilon \le \varepsilon_0$, where ε_0 is given in Theorem 1.

Likewise, one proves, in the case of the *periodic boundary conditions (2.2)*, the following result.

COROLLARY 3.3. *Assume that the hypotheses of Theorem 3.2 hold and that, in addition, one has*

$$(I - M)\mathbf{P}_\varepsilon f = 0,$$

then the attractors $\mathcal{A}_\varepsilon^*$ and $\mathcal{A}_0(g)$ coincide for $0 < \varepsilon \le \varepsilon_0^*$, where ε_0^* is given in Theorem 2.

3.4 Upper-semicontinuity result. If $(I - M)\mathbf{P}_\varepsilon f \ne 0$, the comparison of the two attractors $\mathcal{A}_\varepsilon$ (resp. $\mathcal{A}_\varepsilon^*$) and $\mathcal{A}_0(g)$ is more difficult. However we have obtained the upper-semicontinuity of the attractors $\mathcal{A}_\varepsilon$ (resp. $\mathcal{A}_\varepsilon^*$) at $\varepsilon = 0$.

Let us consider a sequence of positive numbers $\varepsilon_n \to 0$ when $n \to \infty$. We introduce a sequence of functions f_n in $\mathbf{L}^2(Q_3)$ and we set $g_n = \mathbf{P}_{\varepsilon_n} f_n$. In [Raugel and Sell, 1990a], we have shown the following result.

THEOREM 3.4. *(i) Let η_i, $i = 1, 2, 3, 4, r, p$ satisfy Hypothesis $H(a, b)$, where a and b are sufficiently large and assume that the condition*

$$(3.14) \qquad \varepsilon^{4+2r}\eta_4^{-4}(\varepsilon) \to 0 \text{ as } \varepsilon \to 0^+,$$

holds. Let f_n be a sequence in $\mathbf{L}^2(Q_3)$ satisfying

$$(3.15) \qquad \lim_{n \to \infty} \|M\mathbf{P}_{\varepsilon_n} f_n - g_0\| = 0,$$

for some function g_0 in $H_0 \equiv MH_\varepsilon(Q_3)$. Assume further that

$$\|M\mathbf{P}_{\varepsilon_n} f_n\|^2 \le \eta_2^{-2}, \quad \|(I - M)\mathbf{P}_{\varepsilon_n} f_n\|^2 \le \varepsilon_n^r \eta_4^{-2}.$$

Then the attractors $\mathcal{A}_{\varepsilon_n}$ of (2.13), with forcing term $\mathbf{P}_{\varepsilon_n} f_n$, are upper-semicontinuous at $\varepsilon = 0$ in $V_{\varepsilon_n}^1$, i.e.,

$$(3.16) \qquad \sup_{u_n \in \mathcal{A}_{\varepsilon_n}} \inf_{v \in \mathcal{A}_0(g_0)} \|A_{\varepsilon_n}^{1/2}(u_n - v)\| \to 0 \text{ as } \varepsilon_n \to 0$$

where $\mathcal{A}_0(g_0)$ is the global attractor of (2.24).

(ii) If, moreover,

(3.17)
$$\lim_{n \to \infty} \|f_n - f_0\| = 0,$$

where $f_0 \in L^2(Q_2)$, then the attractors $\mathcal{A}_{\varepsilon_n}$ are upper semicontinuous at $\varepsilon = 0$ in $V_{\varepsilon_n}^2$, i.e.,

(3.18)
$$\sup_{u_n \in \mathcal{A}_{\varepsilon_n}} \inf_{v \in \mathcal{A}_0(g_0)} \|A_{\varepsilon_n}(u_n - v)\| \to 0 \text{ as } \varepsilon_n \to 0.$$

Likewise, in the special case of the *periodic boundary conditions (2.2)*, we obtain the following upper-semicontinuity result the proof of which is given in [Raugel and Sell, 1990b].

THEOREM 3.5. *Assume that the Navier-Stokes equations (2.1) are supplemented with the periodic boundary conditions (2.2). Let η_i, $i = 1, 2, 3, 4$, $p < -1, q_1 < 0, q_2 < 0, r < 0$ satisfy Hypothesis 1. Let f_n be a sequence in $L^2(Q_3)$ satisfying the condition (3.15) as well as*

$$\|M P_{\varepsilon_n} f_n\|^2 \leq \varepsilon_n^{q_2} \eta_2^{-2}, \quad \|(I - M) P_{\varepsilon_n} f_n\|^2 \leq \varepsilon_n^r \eta_4^{-2}.$$

Then the attractors $\mathcal{A}_{\varepsilon_n}^*$ of (2.13), with forcing term $P_{\varepsilon_n} f_n$, are upper-semicontinuous at $\varepsilon = 0$ in $V_{\varepsilon_n}^1$, i.e., the property (3.16) holds. If, moreover, the condition (3.17) is satisfied, then the attractors $\mathcal{A}_{\varepsilon_n}^*$ of (2.13) are upper-semicontinuous at $\varepsilon = 0$ in $V_{\varepsilon_n}^2$, i.e., the property (3.18) holds.

3.5 Remarks about the case of homogeneous Dirichlet boundary conditions (2.4). This case is quite different from the cases previously studied. Here we assume that Q_2 is a smooth bounded domain in $\mathbf{R}^2$ and we introduce the spaces

$$H_\varepsilon = \{u \in L^2(Q_3) : \nabla_\varepsilon \cdot u = 0, u \cdot n|_{\partial Q_\varepsilon} = 0\}$$

and

$$V_\varepsilon^1 = \{u \in H_0^1(Q_3) : \nabla_\varepsilon \cdot u = 0\}$$

where n is the outer normal to the boundary ∂Q_3. We let denote by P_ε the orthogonal projection of $L^2(Q_3)$ onto H_ε. Applying P_ε to (2.10), we obtain the dilated Navier-Stokes evolutionary equation (2.13), where $u = P_\varepsilon u \in H_\varepsilon$, $A_\varepsilon u = -P_\varepsilon \Delta_\varepsilon u$ (with homogeneous Dirichlet boundary conditions). One has $V_\varepsilon^1 = D(A_\varepsilon^{1/2})$ and, if we set $V_\varepsilon^2 = D(A_\varepsilon)$ and if we use the regularity results given in Dauge (1984, 1989), one obtains that $V_\varepsilon^2 = V_\varepsilon^1 \cap (H^2(Q_3))^3$. Using the classical Poincare inequality, one shows at once that the estimates (2.15) still hold. Arguing as in Hale and Raugel (1989a, Corollary 2.8), one shows that

(3.19)
$$\|A_\varepsilon^i u\| \leq C_4 \|A_\varepsilon^{i+1/2} u\|, \text{ for } i = 0, 1,$$

where C_4 is a positive constant independent of ε. Using the inequality (3.19) several times and the regularity results of Dauge, one proves that

(3.20)
$$(i) \quad C_5(\|u\|_{H^2(Q)} + \varepsilon^{-1}\|D_3 u\|_{H^1(Q_3)} + \varepsilon^{-2}\|D_3^2 u\|) \leq \varepsilon^{-1}\|A_\varepsilon u\|$$

and

(3.20) (ii) $\|A_\varepsilon u\| \le C_6(\|u\|_{H^2(Q_3)} + \varepsilon^{-1}\|D_3 u\|_{H^1(Q_3)} + \varepsilon^{-2}\|D_3^2 u\|).$

In the case of the boundary conditions (2.4), we do not use the decomposition $u = Mu + (I - M)u$. In Raugel and Sell (1990a), we have proved the following regularity result (where $0 < \varepsilon \le 1$).

THEOREM 3.6. *Let p and r be two real numbers satisfying $-1 < p < 0$ and $r > -3$ and let $\widetilde{C}_1$ and $\widetilde{C}_2$ be two positive constants. Then there exists a positive real number ε_0 such that, for $0 < \varepsilon \le \varepsilon_0$, whenever $u_0 \in D(A_\varepsilon^{1/2})$, $f \in L^\infty((0,\infty), L^2(Q_3))$ satisfy*

$$\|A_\varepsilon^{1/2} u_0\|^2 \le \widetilde{C}_1 \varepsilon^p, \quad \|\mathbf{P}_\varepsilon f\|_\infty^2 \le \widetilde{C}_2 \varepsilon^r,$$

then (2.13) has a (unique) strong solution u that belongs to $C^0([0,\infty), V_\varepsilon^1) \cap L^\infty([0,\infty), V_\varepsilon^1)$ and, for $t \ge 0$,

$$\|A_\varepsilon^{1/2} u(t)\|^2 \le \exp\left(-\frac{\nu C_4^{-2}\varepsilon^{-2} t}{2}\right) \widetilde{C}_1 \varepsilon^p + 2C_4^2 \widetilde{C}_2 \nu^{-2}\varepsilon^{2+r}.$$

Assume now that the forcing term f *is independent of* t. Let $S_\varepsilon(\mathbf{P}_\varepsilon f, t)u_0$ denote the strong solution of (2.13) with initial data u_0 in $\mathcal{B}_{\varepsilon,1}$ where $\mathcal{B}_{\varepsilon,1} = \{u \in V_\varepsilon^1 : \|A_\varepsilon^{1/2} u\|^2 \le \widetilde{C}_1 \varepsilon^p + 2C_4^2 \widetilde{C}_2 \nu^{-2}\varepsilon^{2+r}\}$. In Raugel and Sell (1990a), we have proved the following results concerning the attractor $\mathcal{A}_\varepsilon$ of the equation (2.13). We set: $\mathcal{B}_{\varepsilon,2} = \bigcup_{t\ge 0} S_\varepsilon(\mathbf{P}_\varepsilon f, t)\mathcal{B}_{\varepsilon,1}$.

THEOREM 3.7. *Assume that the hypotheses of Theorem 3.6 hold and that f is independent of t. Let $\varepsilon_0 > 0$ be given in Theorem 3.6. Then, for $0 < \varepsilon \le \varepsilon_0$, the semigroup $S_\varepsilon(\mathbf{P}_\varepsilon f, t)$ generated by the strong solutions of (2.13) has a unique, maximal compact (local) attractor $\mathcal{A}_\varepsilon$, included in $\mathcal{B}_{\varepsilon,2}$, which attracts $\mathcal{B}_{\varepsilon,2}$ in the space V_ε^1. Furthermore*

$$\mathcal{A}_\varepsilon \subset \left\{ u \in V_\varepsilon^1 : \|A_\varepsilon^{1/2} u\|^2 \le 2C_4^2 \widetilde{C}_2 \nu^{-2}\varepsilon^{2+r} \right\}.$$

Moreover, $\mathcal{A}_\varepsilon$ is bounded and compact in V_ε^2, and attracts the bounded set $\mathcal{B}_{\varepsilon,2} \cap V_\varepsilon^2$ in the space V_ε^2.

Finally, the attractor $\mathcal{A}_\varepsilon$ is the global attractor for the weak (Leray) solutions of (2.13).

We have the following result concerning the "size" of $\mathcal{A}_\varepsilon$.

PROPOSITION 3.8. *Assume that the hypotheses of Theorem 3.7 hold. Then*

$$\sup_{u \in \mathcal{A}_\varepsilon} \|u\| \to 0 \text{ as } \varepsilon \to 0.$$

If in addition $r > -2$, then we have the following better convergence result,

$$\sup_{u \in \mathcal{A}_\varepsilon} \|A_\varepsilon^{1/2} u\| \to 0 \text{ as } \varepsilon \to 0.$$

REFERENCES

A. V. BABIN AND M. I. VISHIK, *Attractors of partial differential evolution equations and their dimension*, Russian Math. Survey, 38 (1983), pp. 151–213.

A. V. BABIN AND M. I. VISHIK, *Attractors of evolutionary equations (in Russian)*, Nauka (1989).

L. CAFFARELLI, R. KOHN AND L. NIRENBERG, *Partial regularity of suitable weak solutions of the Navier-Stokes equations for incompressible viscous fluids.*

P. CONSTANTIN AND C. FOIAS, *Global Lyapunov exponents, Kaplan-Yorke formulas and the dimension of the attractor for the two-dimensional Navier Stokes equations*, Comm. Pure Appl. Math. 38 (1985), pp. 1–27.

P. CONSTANTIN AND C. FOIAS, *Navier-Stokes Equations*, Univ. Chicago Press, Chicago (1988).

P. CONSTANTIN, C. FOIAS, O. MANLEY AND R. TEMAM, *Determining modes and fractal dimension of turbulent flows*, J. Fluid Mech. 150 (1985), pp. 427–440.

P. CONSTANTIN, C. FOIAS AND R. TEMAM, *Attractors representing turbulent flows*, Memoirs of AMS 53, (1985), p. 314.

M. DAUGE, *Régularité et singularités des systèmes de Stokes et Navier-Stokes dans des domaines non réguliers de $\mathbf{R}^2$ ou $\mathbf{R}^3$*, Séminaire d'Equations aux Dérivées Partielles de Nantes (1984).

M. DAUGE, *Stationary Stokes and Navier-Stokes systems on two- or three-dimensional domains with corners, Part I: Linearized equations*, SIAM J. on Math. Anal. (1989).

C. FOIAS, C. GUILLOPÉ AND R. TEMAM, *New and a priori estimates for Navier-Stokes equations in dimension 3*, Comm. PDE, 6 (1981), pp. 329–359.

C. FOIAS, G. R. SELL AND R. TEMAM, *Inertial manifolds for nonlinear evolutionary equations*, J. Differential Equations, 73 (1988), pp. 309–353.

C. FOIAS AND R. TEMAM, *Some analytic and geometric properties of the solutions of the Navier-Stokes equations*, J. Math. Pures Appl., 58 (1979), pp. 339–368.

C. FOIAS AND R. TEMAM, *The connection between the Navier-Stokes equations, dynamical systems, and turbulence theory*, in Directions in Partial Differential Equations, Academic Press, New York (1987), pp. 55–73.

A. FRIEDMAN, *Partial Differential Equations of Parabolic Type*, Prentice-Hall, Englewood Cliffs, N.J. (1964).

H. FUJITA AND T. KATO, *On the Navier-Stokes initial value problem I*, Arch. Rational Mech. Anal., 16 (1964), pp. 269–315.

A. V. FURSIKOV, *On some control problems and results concerning the unique solvability of a mixed boundary value problem for the three-dimensional Navier-Stokes and Euler Systems*, Soviet. Math. Dokl., 21 (1980), pp. 889–893.

Y. GIGA, *Book review*, Bull. Amer. Math. Soc., 19 (1988), pp. 337–340.

J. K. HALE, *Asymptotic Behavior of Dissipative Systems*, Math. Surveys and Monographs, Vol. 25, Amer. Math. Soc., Providence, R.I. (1988).

J. K. HALE AND G. RAUGEL, *Reaction diffusion equation on thin domains*, J. Math. Pures Appl., (to appear in 1992), also preprint of Georgia Institute of Technology (1989a).

J. K. HALE AND G. RAUGEL, *A damped hyperbolic equation on thin domains*, Trans. Amer. Math. Soc. (to appear in 1992), also preprint of Georgia Institute of Technology (1989b).

J. K. HALE AND G. RAUGEL, *Partial differential equations on thin domains*, Proceedings of the international conference in Alabama, March 1990, In Differential Equations and Mathematical Physics (Ed. C. Bennewitz), Academic Press (1991a), pp. 63–98.

J. K. HALE AND G. RAUGEL, *Convergence in gradient-like systems*, Preprint of Georgia Institute of Technology, Atlanta (1991b), to appear in ZAMP (1992).

J. K. HALE AND G. RAUGEL, *Dynamics of partial differential equations on thin domains*, in preparation (1991c).

D. HENRY, *Geometric Theory of Semilinear Parabolic Equations*, Lecture Notes in Mathematics, No. 840, Springer-Verlag, New York (1981).

E. HOPF, *Über die Anfangswertaufgabe für die hydrodynamischen Grundgleichungen*, Math. Nachr., 4 (1951), pp. 213–231.

G. KOMATSU, *Global analyticity up to the boundary of solutions of the Navier-Stokes equation*, Comm. Pure Appl. Math., 33 (1980), pp. 545–566.

O. A. LADYZHENSKAYA, *On the classicality of generalized solutions of the general nonlinear nonstationary Navier-Stokes equations*, Trudy Mat. Inst. Steklov 92 (1966), pp. 100–115.

O. A. LADYZHENSKAYA, *The Mathematical Theory of Viscous Incompressible Flow*, 2nd ed., English translation, Gordon and Breach, New York (1969).

O. A. LADYZHENSKAYA, *On the dynamical system generated by the Navier-Stokes equations*, English translation J. Soviet Math., 3, (1972a), pp. 458–479.

O A. LADYZHENSKAYA, *Dynamical system generated by the Navier-Stokes equations*, Dokl. Akad. Nauk SSSR, 205, No. 2, 1972, pp. 318–320; English translation, Soviet Physics-Doklady, 17, No. 7, 1973, pp. 647–649 (1972b).

J. LERAY, *Etude de diverses équations intégrales nonlinéaires et de quelques problèmes que pose l'hydrodynamique*, J. Math. Pures Appl., 12 (1933), pp. 1–82.

J. LERAY, *Essai sur les mouvements plans d'un liquide visqueux que limitent des parois*, J. Math. Pures Appl., 13 (1934a), pp. 331–418.

J. LERAY, *Sur le mouvement d'un liquide visqueux emplissant l'espace*, Acta. Math., 63 (1934b), pp. 193–248.

J. L. LIONS, *Quelques Méthodes de Résolution des Problèmes aux Limites non linéaires*, Gauthier Villars, Paris (1969).

J. L. LIONS AND G. PRODI, *Un théorème d'existence et d'unicité dans les équations de Navier-Stokes en dimension 2*, C.R. Acad. Sc. Paris, 248 (1959), pp. 3519–3521.

J. MALLET-PARET, *Negatively invariant sets of compact maps and an extension of a theorem of Cartwright*, J. Differential Equations, 22 (1976), pp. 331–348.

K. MASUDA, *On the analyticity and unique continuation theorem for solutions of the Navier Stokes equation*, Proc. Japan Acad., 43 (1967), pp. 827–832.

G. RAUGEL, *Continuity of attractors*, Math. Modeling and Numer. Anal., M2AN, 23 (1989), pp. 519–533.

G. RAUGEL AND G. R. SELL, *Equations de Navier-Stokes dans des domaines minces en dimension trois: régularité globale*, C.R. Acad. Sci. Paris, 309 (1989), pp. 299–303.

G. RAUGEL AND G. SELL (1990a), *Navier-Stokes equations in thin three-dimensional domains I: Global attractors and global regularity of solutions*, to appear in J. Amer. Math. Soc.

G. RAUGEL AND G. SELL (1990b), *Navier-Stokes equations in thin three-dimensional domains II global regularity of spatially periodic solutions*, to appear in Séminaire du Collège de France, 1990-1991, Pitmann.

R. J. SACKER AND G. R. SELL, *Lifting properties in skew-product flows with applications to differential equations*, Memoirs Amer. Math. Soc., No. 190 (1977).

R. J. SACKER AND G. R. SELL, *Dichotomies for linear evolutionary equations in Banach spaces*, Preprint (1990).

G. R. SELL, *Nonautonomous differential equations and topological dynamics I: The basic theory*, Trans. Amer. Math. Soc., 127 (1967a), pp. 241–262.

G. R. SELL, *Nonautonomous differential equations and topological dynamics II: Limiting equations*, Trans. Amer. Math. Soc., 127 (1967b), pp. 263–283.

G. R. SELL, *Differential equations without uniqueness and classical topological dynamics*, J. Differential Equations, 14 (1973), pp. 42–56.

J. SERRIN, *On the interior regularity of weak solutions of the Navier-Stokes equations*, Arch. Rational Mech. Anal., 9 (1962), pp. 187–195.

R. TEMAM, *Navier-Stokes Equations*, North-Holland, Amsterdam (1977).

R. TEMAM, *Navier-Stokes Equations and Nonlinear Functional Analysis*, CBMS Regional Conference Series, No. 41, SIAM, Philadelphia (1983).

R. TEMAM, *Infinite Dimensional Dynamical Systems in Mechanics and Physics*, Springer-Verlag, New York, pp. 1988.

W. VON WAHL, *The Equations of Navier-Stokes and Abstract Parabolic Problems*, Vieweg and Sohn, Braunschweig (1985).

AN OPTIMALITY CONDITION FOR
APPROXIMATE INERTIAL MANIFOLDS*

GEORGE R. SELL†

Abstract. In this paper, we present a basic theoretical framework for interpreting essentially all known methods for constructing approximate inertial manifolds for any of a large class of nonlinear evolutionary equations. This class includes the 2D Navier-Stokes equations and many systems of reaction diffusion equations. We prove that, under reasonable assumptions, every approximate inertial manifold for a given equation is an actual inertial manifold of an approximate equation. This new theoretical framework allows one to introduce certain optimality conditions for approximate inertial manifolds. Finally we introduce a new method, the Gamma Method, for the construction of approximate inertial manifolds.

Key words. approximate inertial manifolds, attractors, inertial manifolds, Navier-Stokes equations, reaction diffusion equations

1. Introduction. The theory of inertial manifolds is an important development in the study of dissipative dynamical systems on infinite dimensional Hilbert spaces. The infinite dimensional system is typically given as the nonlinear semigroup generated by the solutions of a nonlinear partial differential equation (PDE), which in turn is oftentimes described as an abstract nonlinear evolutionary equation,

$$(1.1) \qquad u' + Au = F(u), \qquad u(0) = u_0,$$

on some Hilbert space $\mathcal{H}$. When such an infinite dimensional system has an inertial manifold $\mathfrak{M}$, then the long-time dynamics is completely described by the solutions of a finite system of ordinary differential equations (ODE). This ODE is referred to as an inertial form for the PDE. (See, for example, Foias, Sell, and Temam (1988). Additional references are given in Mallet-Paret and Sell (1988).)

One of the properties of the inertial manifold is that it attracts all solutions of the underlying PDE at an exponential rate. In some cases, this exponential rate of convergence is very large. Heuristically speaking, given any solution of the PDE, there is a time τ, which depends on the initial value of the solution, such that the given solution appears to be *on* the inertial manifold for $t > \tau$. Since the time τ can be very small, depending on how close the initial condition is to the inertial manifold, this implies that, in addition to the long-time dynamics, some of the transient dynamics of solutions of the PDE are described by solutions of the inertial form.

The theory of inertial manifolds has special significance in the numerical computation of solutions of the underlying PDE. In order to do any computation of a solution of a PDE, the first step is to replace the PDE with some finite dimensional approximation, for example a Galerkin approximation. This step in going from the

---

*This research was supported in part by grants from the National Science Foundation, the Applied Mathematics and Computational Mathematics Program/DARPA, and the U. S. Army.

†School of Mathematics, University of Minnesota, Minneapolis, Minnesota 55455

infinite dimensional problem to a finite dimensional problem is normally a fundamental cause of error in any numerical computation. In the case where the PDE has an inertial form, then the reduction to the finite dimensional problem is exact; it introduces **NO ERROR**. Therefore, approximation schemes based on the theory of inertial manifolds hold the promise for significant increases in computational efficiency.

In order to exploit the promise of increased computational efficiency, it is important to find a good theory and good algorithms for approximating the inertial manifold and the associated inertial form. Furthermore one would like to develop a theory for an approximate inertial manifold (AIM) which would be useful for both dynamical studies and computational purposes.

Most theories of approximate inertial manifolds require that the global attractor $\mathfrak{A}$ for the underlying PDE be *close to* the AIM. The topology which is oftentimes used to describe *close to* is the C^0-topology. The requirement that $\mathfrak{A}$ be close to the AIM in the C^0-topology may lead to some information about the dynamics on the global attractor. For example, some theories do give good information about the fixed points, or stationary solutions of the underlying PDE. However, a C^0-approximation is usually too crude to be useful for a reliable understanding of the more delicate dynamical features such as periodic solutions or chaotic attractors. What one needs is a C^1-approximation instead. One wants both the global attractor $\mathfrak{A}$ to be close to the AIM, *and* the vector field on it to be close to the associated approximate inertial form.

In this paper we will examine a class of nonlinear evolutionary equations (1.1), a class which includes the Navier-Stokes equations on a sufficiently smooth bounded domain in R^2, as well as many systems of reaction diffusion equations. In this context, we wish to briefly review the current literature on AIMs for these equations. Our first objective in this paper is to prove that, under reasonable conditions, **every approximate inertial manifold for a given equation is an actual inertial manifold of an approximate equation,**

(1.2)
$$u' + Au = F(u) + H(u),$$

where the perturbation term H is known. See Section 3.

It turns out that essentially all methods realize the AIM as the graph of a suitable function Φ, which lies in a Banach space $W^{1,\infty}$. A convenient measure of how close the AIM for (1.1) is to an actual inertial manifold is given in terms of a suitable norm $\|H\|$ on the perturbation term H, i.e., $\mathfrak{M} = \text{graph } \Phi$ is a inertial manifold for (1.1) if and only if $\|H\| = 0$. The optimality condition referred to in the title is to minimize the norm $\|H\|$ over suitable subsets of $W^{1,\infty}$. This is described in Section 4.

In Section 5 we reexamine some of the AIM methods for the purpose of calculating the C^0 and C^1 norms of the related perturbation terms $\|H\|$. As noted in Pliss and Sell (1991), this forms the basis for some useful information concerning the study of the Approximation Dynamics of (1.1).

Finally, we present a new general method for the construction of AIMs. This method, which we call the **Gamma Method**, is presented in Sections 6 and 7. We show that the Gamma Method has wide applicability. In particular, we show that the Navier-Stokes equations on s suitable bounded region in R^2 have AIMs of arbitrarily small order. This holds for any value of the viscosity, or Reynolds number. We also show that all dissipative systems of reaction diffusion equations have AIMs of arbitrarily small order in any space dimension. In the case where (1.1) does admit an inertial manifold, the Gamma Method offers a special advantage over all other methods, because by using it one can find good AIMs of *lower* dimension than the inertial manifold.

2. Attractors. We will be studying the dynamics of infinite dimensional non-linear evolutionary equation of the form

$$(2.1) \qquad u' + Au = F(u), \qquad u(0) = u_0,$$

where u_0 is some initial condition in the Hilbert space $\mathcal{H}$. We assume that A is a self-adjoint, positive, linear operator with compact resolvent and domain $\mathcal{D}(A) \subset \mathcal{H}$. The term F contains the nonlinear terms. We assume that there is an β, $\frac{1}{2} \leq \beta \leq 1$, such that

$$F : \mathcal{D}(A) \to \mathcal{D}(A^\beta)$$

is a C^1-mapping (in the sense of Gateaux) in the respective norms on the spaces $\mathcal{D}(A)$ and $\mathcal{D}(A^\beta)$, and the derivative DF is uniformly bounded on bounded sets in $\mathcal{D}(A)$. We also assume that (2.1) has a unique well-posed solution $S(t)u_0$ for every $u_0 \in \mathcal{H}$ and that $S(t)u_0 \in \mathcal{D}(A)$ for all $t > 0$. This occurs, for example, in the case of the 2D Navier-Stokes equations on a suitable bounded domain, or in the case that $F : \mathcal{H} \to \mathcal{D}(A^{\beta-1})$ is locally Lipschitz continuous, where $\mathcal{D}(A^{\beta-1})$ is the dual space of $\mathcal{D}(A^{\beta-1})$, see Temam (1977, 1983, 1988a).

We assume further that equation (2.1) is **locally dissipative**. By this we mean that there is an open set $\mathcal{O}$ in $\mathcal{D}(A)$ and a closed, bounded set $\mathcal{B}$ in $\mathcal{D}(A)$ satisfying the following properties:

$$\mathcal{B} \subset \mathcal{O}, \qquad S(t)\mathcal{O} \subset \mathcal{O} \quad \text{for } t \geq 0,$$

and for every $u_0 \in \mathcal{O}$ there is a time $T = T(u_0)$ such that

$$S(t)u_0 \in \mathcal{B} \qquad \text{for } t \geq T.$$

The local dissipative property implies that (2.1) has an attractor[1] $\mathfrak{A}$ in $\mathcal{B}$, and that the basin of attraction of $\mathfrak{A}$ contains the open set $\mathcal{O}$. It is known that this attractor is nonempty and has finite Hausdorff dimension, see Hale (1988) and Temam (1988). If it happens that the basin of attraction of $\mathfrak{A}$ is the entire space $\mathcal{D}(A)$, then the attractor $\mathfrak{A}$ is referred to as the **global attractor** for (2.1).

---

[1] We use the word **attractor** in this paper to refer to a compact, attracting, invariant set that is Lyapunov stable, see Hale (1988).

Since we are interested in the long-time dynamics of the solutions of (2.1), and since we are assuming (2.1) to be locally dissipative, there is no loss in generality in modifying either the AIM $\mathfrak{M}_a$, or the nonlinearity in equation (2.1) near ∞. For example, if one prepares equation (2.1), as in Foias, Sell, and Temam (1988), then one could restrict to AIMs $\mathfrak{M}_a = \text{graph } \Phi_a$ where Φ_a is bounded and globally Lipschitz continuous. Alternatively, one could, if necessary, modify Φ_a near $|p| = \infty$ to insure that Φ_a is bounded and globally Lipschitz continuous. In other words, there is no loss in generality in assuming Φ_a to lie in the Sobolev space $W^{1,\infty}$, where

$$W^{1,\infty} \overset{\text{def}}{=} W^{1,\infty}(P\mathcal{H}, Q\mathcal{H} \cap \mathcal{D}(A)).$$

To be specific, we will assume further that the nonlinearity $F = F(u)$ in (2.1) satisfies the following condition: There is a $\rho > 0$ such that

$$(2.2) \qquad\qquad F(u) = 0, \qquad \text{for } |Au| \geq \rho.$$

It follows from (2.2) and the other assumptions on F, that there are constants C_0 and C_1 such that

$$(2.3) \qquad\qquad |A^\beta F(u)| \leq C_0$$

for all $u \in \mathcal{D}(A)$, and

$$(2.4) \qquad\qquad |A^\beta DF(u)v| \leq C_1 |Av|$$

for all $u, v \in \mathcal{D}(A)$. Notice that (2.4) implies that one has

$$(2.5) \qquad\qquad |A^\beta (F(u_1) - F(u_2))| \leq C_1 |Au_1 - Au_2|$$

for all $u_1, u_2 \in \mathcal{D}(A)$.

We shall say that (2.1) satisfies the **Standing Hypotheses** when the conditions[2] stated in the last three paragraphs hold. It can happen that a given flow on $\mathcal{H}$ has many local attractors. (The global attractor, when it exists, is of course unique.) For the theory presented below we shall treat $\mathfrak{A}$ as a distinguished attractor. Our objective is to approximate well the dynamics on $\mathfrak{A}$.

A typical problem that gives rise to a nonlinear evolutionary equation satisfying the above hypotheses is the Navier-Stokes equations on a minimally smooth bounded region in R^2 (see Constantin and Foias (1988), Ladyzhenskaya (1972) and Temam (1977, 1983, 1988)), or on a suitable thin bounded regions in R^3 (see Raugel and Sell (1990)). In this setting one has $\beta = \frac{1}{2}$, see Foias, Sell, and Temam (1988). In the R^2 case, there is a global attractor for the Navier-Stokes equations. As a

---

[2] There are, of course, other formulations of the theory of the long-time dynamics of nonlinear evolutionary equations. For example, if one studies the strong solutions of the 3D Navier-Stokes equations, then it is appropriate to take the initial condition u_0 in $\mathcal{D}(A^{\frac{1}{2}})$, see Constantin and Foias (1988) and Temam (1977, 1983). Similarly, for some systems of reaction diffusion equations, it is more common to restrict to $u_0 \in \mathcal{D}(A^{\frac{1}{2}})$ as well, cf. Henry (1981).

matter of fact, in the R^2 setting, the basin of attraction is the full space $\mathcal{H}$, which is typical for many nonlinear evolutionary equations with global attractors. For the Navier-Stokes equations in thin regions in R^3, there is an attractor with a large basin of attraction in $\mathcal{D}(A)$. Furthermore, this attractor turns out to be a global attractor for the weak solutions, see Raugel and Sell (1990). This general formulation (with $\beta = 1$) includes many systems of reaction diffusion equations, see Henry (1981).

The eigenvalues of A will be denoted by

$$0 < \lambda_1 \le \lambda_2 \le \lambda_3 \le \ldots \lambda_n \le \ldots,$$

and the corresponding eigenvectors are given by

$$e_1, e_2, e_3, \ldots, e_n, \ldots.$$

For $n \ge 1$ we define the projection $P = P_n$ to be the orthogonal projection onto

$$\text{span } \{e_1, e_2, \ldots, e_n\},$$

and we let $Q = Q_n = I - P$ be the complementary projection. The projections P and Q commute with A. If we apply both P and Q to (2.1) we get the equivalent system

(2.6) $$p' + APp = PF(p+q), \qquad q' + AQq = QF(p+q)$$

where $p = Pu$ and $q = Qu$.

The concept of an inertial manifold was formulated in order to reduce the study of the dynamics of the *global* attractor of (2.1) to an ODE, see Foias, Sell and Temam (1988). The same ideas apply to any attractor $\mathfrak{A}$. Recall that a subset $\mathfrak{M}$ of $\mathcal{H}$ is an **inertial manifold** (for $\mathfrak{A}$) if the following properties are satisfied:

(1) $\mathfrak{M}$ is a smooth, finite dimensional manifold.

(2) $\mathfrak{M}$ is positively invariant under the flow $S(t)$, i.e., $S(t)\mathfrak{M} \subset \mathfrak{M}$ for $t \ge 0$.

(3) $\mathfrak{M}$ is **locally exponentially attracting**, i.e., there is an $\nu > 0$ such that for every $u_0 \in \mathcal{O}$ there is a $K = K(u_0)$ such that

(2.7) $$\text{dist}_{\mathcal{H}}(S(t)u_0, \mathfrak{M}) \le Ke^{-\nu t}, \qquad t \ge 0.$$

Since $\mathcal{O}$ is positively invariant, the local exponential attracting property implies that

$$\mathfrak{A} \subset \mathfrak{M} \cap \mathcal{O} \subset \mathcal{O}.$$

The concept of local exponential attracting does arise elsewhere, see Mallet-Paret and Sell (1988) for example. If (2.7) is holds for every $u_0 \in \mathcal{H}$, then $\mathfrak{M}$ is said to be **globally exponentially attracting**. If $\mathfrak{A}$ is a global attractor, then the neighborhood $\mathcal{O}$ is usually chosen to be the entire space $\mathcal{D}(A)$, or where appropriate, $\mathcal{H}$, see Foias, Sell and Temam (1988), for example.

We are especially interested in the case where $\mathfrak{M}$ can be realized as the graph of a smooth function Φ from $P\mathcal{H}$ to $Q\mathcal{H} \cap \mathcal{D}(A)$. Note that in this case, $\mathfrak{M} = $ graph Φ is positively invariant if and only if the following is valid: The function $q(t) = \Phi(p(t))$ is a solution of

$$(2.8) \qquad q' + AQq = QF(p(t) + q)$$

whenever $p(t)$ is a solution of

$$(2.9) \qquad p' + APp = PF(p + \Phi(p)).$$

Since Φ is a smooth function, one can calculate $q' = \frac{\partial}{\partial t} q$ by means of the chain rule:

$$q' = D\Phi\, p',$$

where $D\Phi$ denotes the derivative of Φ with respect to p. From (2.8) and (2.9) we then get

$$(2.10) \qquad D\Phi(PF(p + \Phi) - APp) = QF(p + \Phi) - AQ\Phi.$$

In other words, the manifold $\mathfrak{M} = $ graph Φ, where Φ is a C^1-function, is positively invariant for (2.6) if and only if (2.10) holds. (See Fabes, Luskin, and Sell (1988) for another characterization of positive invariance.)

3. Approximate inertial manifolds. The theory of inertial manifolds offers a new paradigm for the study of the long-time dynamics of the solutions of certain PDEs. Since the existence of the associated inertial form gives an exact reduction of an infinite dimensional problem to a finite dimensional problem, approximation schemes based on the theory of inertial manifolds hold the promise for significant increases in computational efficiency. Current efforts in developing new AIM methods illustrate the inertial manifold paradigm.

As we shall see below, essentially all AIM methods can be reduced to the following scheme: First one selects a projection P with finite dimensional range and applies P and $Q = I - P$ to (2.1) to obtain the equivalent system[3]

$$(3.1) \qquad p' + APp = PF(p + q), \qquad q' + AQq = QF(p + q)$$

where $p = Pu$ and $q = Qu$. An AIM $\mathfrak{M}_a$ for (3.1) is then constructed as the graph of some function Φ_a from $P\mathcal{H}$ to $Q\mathcal{H} \cap \mathcal{D}(A)$, i.e., one has

$$\mathfrak{M}_a = \text{graph } \Phi_a = \{p + \Phi_a(p) : p \in P\mathcal{H}\}.$$

---

[3] In order to simplify the treatment, we will assume in this paper that P is an orthogonal projection on $\mathcal{H}$ and that P commutes with A.

Let $\Phi \in W^{1,\infty}$. Our next objective is to show that $\mathfrak{M} = \text{graph } \Phi$ is a positively invariant manifold for some system of the form

$$(3.2) \qquad p' + APp = PF(p+q), \qquad q' + AQq = QF(p+q) + E(p),$$

where the error term E is known and depends on Φ. Indeed, as argued in Section 2, see (2.10), $\mathfrak{M} = \text{graph } \Phi$ is positively invariant for (3.2) if and only if $q' = D\Phi p'$, i.e.,

$$QF(p+\Phi) + E(p) - AQ\Phi = D\Phi \left(PF(p+\Phi) - APp\right).$$

We can solve the last equation for E to obtain

$$(3.3) \qquad E(p) \overset{\text{def}}{=} AQ\Phi - QF(p+\Phi) + D\Phi\left(PF(p+\Phi) - APp\right).$$

In other words, if one defines E by (3.3), then $\mathfrak{M} = \text{graph } \Phi$ is an invariant manifold for (3.1) if and only if $E = 0$.

Since F satisfies (2.2)-(2.5), it follows that if $\Phi \in W^{1,\infty}$, then $E \in L^\infty_{loc}(P\mathcal{H}, Q\mathcal{H})$. If in addition, Φ has compact support, then $E \in L^\infty(P\mathcal{H}, Q\mathcal{H})$.

We next introduce two measures of how far a given AIM $\mathfrak{M} = \text{graph } \Phi$ differs from an actual inertial manifold. We shall refer to these measures as the **defects** of $\mathfrak{M}$, and we shall denote them by Defect_i, $i = 0, 1$. The first is defined as

$$(3.4) \qquad \text{Defect}_0 \mathfrak{M} = \text{Defect}_0(\Phi) = \|E\|_{C^0} \overset{\text{def}}{=} \sup_{|Ap| \leq \rho} |E(p)|.$$

If one has $\Phi \in W^{2,\infty}(P\mathcal{H}, Q\mathcal{H} \cap \mathcal{D}(A))$, then the derivative DE is defined, and Defect_1 is given by

$$\text{Defect}_1 \mathfrak{M} = \text{Defect}_1(\Phi) = \|E\|_{C^1} \overset{\text{def}}{=} \|E\|_{C^0} + \sup_{|Ap| \leq \rho, |A\hat{p}| \leq 1} |DE(p)\hat{p}|.$$

The measure $\text{Defect}_1 \mathfrak{M} = \|E\|_{C^1}$ is a good measure of how well the dynamics on the AIM approximates the dynamics of the original system, see Pliss and Sell (1991).

Our next step is to seek sufficient conditions for the AIM $\mathfrak{M} = \text{graph } \Phi$ to be an actual inertial manifold for the perturbed equation (3.2). The following result addresses this issue:

THEOREM. *Let the Standing Hypotheses hold. Assume that $\lambda^\beta_{n+1} > C_1$, where C_1 is given by (2.4), and let $\Phi \in W^{1,\infty}$. Then the AIM $\mathfrak{M} = \text{graph } \Phi$ is an actual inertial manifold for (3.2) where E is defined by (3.3). In particular, $\mathfrak{M}$ is (globally) exponentially attracting in $\mathcal{D}(A)$. More precisely, if $(p(t), q(t))$ is any solution of (3.2), then one has*

$$(3.5) \qquad |A(q(t) - \Phi(p(t)))|^2 \leq |A(q(0) - \Phi(p(0)))|^2 e^{-\nu t}, \qquad t \geq 0,$$

where $\nu = 2\lambda_{n+1}(1 - C_1 \lambda^{-\beta}_{n+1})$.

Proof. Let $\Phi \in W^{1,\infty}$ and define E by (3.3). We now introduce the change of variables $q = \Phi(p) + r$ into (3.2). Since one has

$$\Phi' + AQ\Phi = QF(p+\Phi) + E(p)$$

and since q satisfies the q-equation in (3.2), we find that (3.2) reduces to

(3.6) $\quad p' + APp = PF(p + \Phi + r), \qquad r' + AQr = QF(p + \Phi + r) - QF(p + \Phi).$

By taking the scalar product of the r-equation in (3.6) with $A^2 r$ one obtains

$$
\frac{1}{2}\frac{d}{dt}|Ar|^2 + |A^{\frac{3}{2}}r|^2 = \langle Q[F(p + \Phi + r) - F(p + \Phi)], A^2 r \rangle
$$
$$
(3.7) \qquad\qquad = \langle A^{\beta}Q[F(p + \Phi + r) - F(p + \Phi)], A^{1+\alpha}r \rangle.
$$

From (2.5) we then obtain

$$
\frac{1}{2}\frac{d}{dt}|Ar|^2 + |A^{\frac{3}{2}}r|^2 \le C_1 |Ar||A^{1+\alpha}r| \le C_1 \lambda_{n+1}^{-\frac{1}{2}}\lambda_{n+1}^{-\frac{1}{2}+\alpha}|A^{\frac{3}{2}}r|^2 \le C_1\lambda_{n+1}^{-\beta}|A^{\frac{3}{2}}r|^2,
$$

which implies that

$$
\frac{1}{2}\frac{d}{dt}|Ar|^2 + \lambda_{n+1}(1 - C_1\lambda_{n+1}^{-\beta})|Ar|^2 \le \frac{1}{2}\frac{d}{dt}|Ar|^2 + (1 - C_1\lambda_{n+1}^{-\beta})|A^{\frac{3}{2}}r|^2 \le 0.
$$

Since $\lambda_{n+1}^{\beta} > C_1$, the Gronwall inequality then implies (3.5). □

REMARKS. 1. What we have shown in the last theorem is that if the dimension of $P\mathcal{H}$ is large enough, in the sense that $\lambda_{n+1}^{\beta} > C_1$, then every approximate inertial manifold for (3.1) is an actual inertial manifold for the approximate equation (3.2).

2. One point worth noting is that while we have proven that the AIM $\mathfrak{M} =$ graph Φ is exponentially attracting for (3.2), we are not claiming that $\mathfrak{M}$ is normally hyperbolic, see Mallet-Paret, Sell, and Shao (1991). As is known, normal hyperbolicity implies that $\mathfrak{M}$ is stable under small perturbations, see Fenichel (1971), Pliss (1977), Sacker (1969), and Pliss and Sell (1991). Normal hyperbolicity for inertial manifolds is typically a consequence of a spectral gap condition, see Mallet-Paret, Sell, and Shao (1991).

4. An optimality condition. Throughout this section we shall fix P and restrict our attention to the case where $\lambda_{n+1}^{\beta} > C_1$. Consequently, the Theorem in Section 3 is applicable here.

Before we describe the optimality condition for approximate inertial manifolds, we shall review some properties of the space $W^{1,\infty}$. First recall that one has

$$
\Phi \in W^{1,\infty} = W^{1,\infty}(P\mathcal{H}, Q\mathcal{H} \cap \mathcal{D}(A))
$$

if and only if Φ is a differentiable mapping (in the sense of distributions) from $P\mathcal{H}$ into $Q\mathcal{H} \cap \mathcal{D}(A)$ and there exist constants K_0 and K_1 such that

$$
(4.1) \quad \begin{cases} |A\Phi(p)| \le K_0, & \text{for all } p \in P\mathcal{H}, \\ |AD\Phi(p)\hat{p}| \le K_1|\hat{p}|, & \text{for almost all } p \in P\mathcal{H} \text{ and all } \hat{p} \in P\mathcal{H}. \end{cases}
$$

The norm of Φ is defined as

$$
|\Phi|_{W^{1,\infty}} \overset{\text{def}}{=} \inf\{K_0 + K_1 : \text{ the inequalities above hold}\}.
$$

Let us return to the operator $\text{Defect}_0(\Phi)$ defined by (3.4). This function is a continuous mapping of $W^{1,\infty}$ into R^+. Since bounded sets in $W^{1,\infty}$ are compact in the weak* topology, it follows that for every bounded set $\mathcal{B} \subset W^{1,\infty}$ there is a $\Psi \in \mathcal{B}$ such that $\text{Defect}_0(\Psi) = e(\mathcal{B})$, where $e(\mathcal{B})$ is defined by

$$e(\mathcal{B}) \overset{\text{def}}{=} \inf_{\Phi \in \mathcal{B}} \text{Defect}_0(\Phi).$$

Since $\lambda_{n+1}^\beta > C_1$, equation (3.2) has an inertial manifold of the form $\mathfrak{M} = \text{graph } \Phi$ if and only if there exists a bounded set $\mathcal{B} \subset W^{1,\infty}$ with $e(\mathcal{B}) = 0$.

The function $\Psi \in \mathcal{B}$ that satisfies $\text{Defect}_0(\Psi) = e(\mathcal{B})$ generates an **optimal** inertial manifold $\mathfrak{M} = \text{graph } \Psi$. However this notion of optimality is a relative concept since it refers to the bounded set $\mathcal{B}$. If $\mathcal{B}_1$ and $\mathcal{B}_2$ are two bounded sets in $W^{1,\infty}$ with $\mathcal{B}_1 \subset \mathcal{B}_2$, then $e(\mathcal{B}_2) \leq e(\mathcal{B}_1)$. What one really would like to do is to prove the existence of an absolute minimum of $\text{Defect}_0(\Phi)$ on $W^{1,\infty}$. In order to do this one needs to develop some *a priori* bounds for $\text{Defect}_0(\Phi)$.

Let us consider the rectangular sets $\mathcal{B}(a,b)$ consisting of all $\Phi \in W^{1,\infty}$ such that $|A\Phi(p)| \leq a$ and $|AD\Phi(p)| \leq b$ for almost all $p \in P\mathcal{H}$ with $|Ap| \leq \rho$. Since $\Phi_0 \equiv 0 \in \mathcal{B}(a,b)$ one has

$$(4.2) \quad e(\mathcal{B}(a,b)) \leq \text{Defect}_0(0) = \sup_{|Ap|\leq\rho} |QF(p)| \leq \lambda_{n+1}^{-\beta} \sup_{|Ap|\leq\rho} |A^\beta F(p)| \leq \lambda_{n+1}^{-\beta} C_0$$

from (2.3). It follows from the definition of E in (3.3), (2.3) and the restriction that $|Ap| \leq \rho$ that for any $\Phi \in \mathcal{B}(a,b)$ one has

$$(4.3)$$
$$|E| \geq |AQ\Phi| - |QF| - |D\Phi||PF| - |D\Phi||Ap| \geq |AQ\Phi| - b\lambda_{n+1}^{-1}\rho - (1 + b\lambda_{n+1}^{-1})\lambda_{n+1}^{-\beta} C_0.$$

Assume next that $|A\Phi| \geq (b\lambda_{n+1}^{-1}\rho + (2 + b\lambda_{n+1}^{-1})\lambda_{n+1}^{-\beta} C_0)$. It then follows from (4.2) and (4.3) that

$$\text{Defect}_0(\Phi) \geq \lambda_{n+1}^{-\beta} C_0 \geq \text{Defect}_0(0).$$

As a result one has

$$e(\mathcal{B}(a,b)) = e(\mathcal{B}(a_0, b)) \leq \lambda_{n+1}^{-\beta} C_0, \qquad \text{for all } a \geq a_0,$$

where $a_0 \overset{\text{def}}{=} (b\lambda_{n+1}^{-1}\rho + (1 + b\lambda_{n+1}^{-1})\lambda_{n+1}^{-\beta} C_0)$.

We have not succeeded in obtaining similar bounds for $e(\mathcal{B}(a,b))$ in terms of the parameter b. Rather than describing some calculations which fall short of the desired result, it seems more useful to reflect on the mathematical properties underlying the difficulties. The basic issue can be summarized as follows:

How can $\text{Defect}_0(\Phi)$ be small while $|AD\Phi|$ is large?

If one has a sequence Φ_n with $\text{Defect}_0(\Phi_n) \to 0$, then the graphs $\mathfrak{M}_n = \text{graph } \Phi_n$ are seeking to approach an inertial manifold. If at the same time one has $|AD\Phi_n| \to \infty$, while $|A\Phi_n|$ remains bounded, then the manifolds $\mathfrak{M}_n$ are developing *shocks*. What this seems to suggest is that the paradigm of seeking an inertial manifold, in this case, as the graph of a function Φ from $P\mathcal{H}$ to $Q\mathcal{H} \cap \mathcal{D}(A)$ is at fault. What this in turn suggests is the desirability of introducing local coordinates to describe both the inertial manifold and the AIM. For related work along these lines, see Foias and Temam (1988).

5. A survey of AIM methods. The theory of approximate inertial manifolds has been studied extensively by a number of authors, including Fabes, Luskin and Sell (1988), Foias, Jolly, Kevrekidis, Sell, and Titi (1988), Foias, Manley, and Temam (1988), Foias, Sell and Titi (1989), Foias and Temam (1988), Luskin and Sell (1988), Marion (1989a, 1989b), Marion and Temam (1989, 1990), Temam (1989), and Titi (1990), as well as the classical Galerkin approximations. The specific issue we address in this section is to determine whether the methods used in these papers do in fact yield AIMs of any defect $\eta > 0$. If so, one would then like to estimate the defect η.

In this section we will describe several methods which have been used to reduce (2.1) to a finite dimensional ODE. We begin with the equivalent system

$$(5.1) \qquad p' + APp = PF(p + q), \qquad q' + AQq = QF(p + q).$$

For each of these methods we shall construct a manifold

$$\mathfrak{M}_i = \text{graph } \Phi_i, \qquad i = 0, 1, 2, \ldots$$

where $\Phi_i : P\mathcal{H} \to Q\mathcal{H} \cap \mathcal{D}(A)$ is a smooth function. We will then argue that $\mathfrak{M}_i$ is the inertial manifold for the perturbed equation

$$(5.2) \qquad p' + APp = PF(p + q), \qquad q' + AQq = QF(p + q) + E_i(p),$$

for an appropriate perturbation term $E_i(p)$ which satisfies $QE_i = E_i$. Finally we shall derive an upper bound for the defect of $\mathfrak{M}_i$ by estimating

$$\sup_{|Ap| \leq \rho} |E_i(p)|, \qquad \text{and} \qquad \sup_{|Ap| \leq \rho} |DE_i(p)|.$$

The approximate inertial form, which is generated by (5.2), is an ODE

$$p' + APp = PF(p + \Phi_i(p)).$$

We assume in this section that (5.1) satisfies the Standing Hypotheses, and that $\lambda_{n+1}^\beta > C_1$. Therefore the Theorem in Section 3 is applicable. The first method we examine is the classical Galerkin approximation.

5.1 Classical Galerkin method. The approach here is to set $q = \Phi_0(p) \stackrel{\text{def}}{=} 0$ in (5.1). The AIM then becomes the flat manifold $\mathfrak{M}_0 = P\mathcal{H}$, and the associated approximate inertial form is

$$(5.3) \qquad p' + APp = PF(p).$$

From Section 3 we see that $\mathfrak{M}_0$ is an inertial manifold for the perturbed system (5.2), with $i = 0$, where

$$E_0(p) = -QF(p).$$

By using (2.3) and (2.5) one then obtains

$$(5.4) \qquad \begin{aligned} |E_0(p)| &\leq \lambda_{n+1}^{-\beta} |A^\beta QF(p)| \leq \lambda_{n+1}^{-\beta} C_0 \\ |DE_0(p)h| &\leq \lambda_{n+1}^{-\beta} |A^\beta QDF(p)h| \leq \lambda_{n+1}^{-\beta} C_1 |Ah| \end{aligned}$$

for all $p \in P\mathcal{H}$. Since $\mathfrak{M}_0$ is a linear space in $\mathcal{H}$, the classical Galerkin methods are sometimes referred to as *linear* methods, even though the approximate inertial form (5.3) is not linear. The next construction involves nonlinear Galerkin methods.

5.2 Nonlinear Galerkin methods. For these methods one begins by setting $q' = 0$ in (5.1). This then leads to the nonlinear equation $q = \Phi(p)$ where $AQ\Phi = QF(p + \Phi)$, or equivalently,

$$\text{(5.5)} \qquad\qquad \Phi = (AQ)^{-1}QF(p + \Phi).$$

Define ν by

$$\nu = \lambda_{n+1}^{-\beta}C_1,$$

where β and C_1 are given by the Standing Hypotheses and (2.4). Since one has

$$\begin{aligned}
\big|A(AQ)^{-1}[QF(p + &\Phi_1(p)) - QF(p + \Phi_2(p))]\big| \\
&= \big|(AQ)^{-\beta}A^\beta[F(p + \Phi_1(p)) - F(p + \Phi_2(p))]\big| \\
&\leq \nu|A\Phi_1(p) - A\Phi_2(p)|,
\end{aligned}$$

the Contraction Mapping Theorem implies that (5.5) has a unique solution Φ_∞ since $\nu < 1$. Furthermore, any sequence of successive approximations converges to Φ_∞. One class of nonlinear Galerkin methods is given by the terms in such a sequence, i.e., set

$$\text{(5.6)} \qquad\qquad \Phi_i(p) \stackrel{\text{def}}{=} (AQ)^{-1}QF(p + \Phi_{i-1}(p)), \qquad 1 \leq i,$$

where $\Phi_0 \equiv 0$. The first approximation Φ_1 in (5.6) is discussed in Foias, Manley, and Temam (1988), Marion (1989a, 1989b), Marion and Temam (1989, 1990), Temam (1989), and Titi (1990), while the second approximation Φ_2 can be found in Temam (1989). Each of the manifolds $\mathfrak{M}_i = $ graph Φ_i is an inertial manifold for (5.2) where

$$E_i \stackrel{\text{def}}{=} D\Phi_i[PF(p + \Phi_i) - APp] + QF(p + \Phi_{i-1}) - QF(p + \Phi_i), \qquad i = 1, 2, \ldots.$$

We now have the following result, which is applicable to the 2D Navier-Stokes equations:

THEOREM. *Assume that the Standing Hypotheses are satisfied. Assume futher that $\lambda_n \geq 1$, $\nu = \lambda_{n+1}^{-\beta}C_1 \leq \frac{1}{2}$, and that $F : \mathcal{D}(A) \to \mathcal{D}(A^\beta)$ is a C^2-function that satisfies*

$$\text{(5.7)} \qquad\qquad |A^\beta D^2 F(u)vw| \leq C_2|Av||Aw|, \qquad u, v, w \in \mathcal{D}(A).$$

Then there are constants C_3 and C_4, which do not depend on i, such that

$$\text{(5.8)} \qquad\qquad |E_i(p)| \leq \lambda_{n+1}^{-\beta}C_3,$$

$$\text{(5.9)} \qquad\qquad |DE_i(p)v| \leq \lambda_{n+1}^{-\beta}C_4|Av|,$$

for all $i \geq 1$, $p \in P\mathcal{H}$ with $|Ap| \leq \rho$, and $v \in \mathcal{D}(A)$.

Proof. The proof of (5.8) and (5.9) comes from a straight forward, but rather lengthy computation. We shall indicate here the main steps and omit many of the details. Except where noted, the inequalities described below hold for all $i \geq 1$ and all $p \in P\mathcal{H}$. the terms $C_5, C_6, \ldots$ appearing here are constants which do not depend on i.

First note that (2.5) implies that

$$|A\Phi_i(p)| = |(AQ)^{-\beta} A^\beta F(p + \Phi_{i-1})| \leq \lambda_{n+1}^{-\beta} C_0.$$

By using (2.5), (5.6), and induction, one finds that

$$|A\Phi_i(p) - A\Phi_{i-1}(p)| \leq \nu^i C_5,$$

where $C_5 = C_0 C_1^{-1}$. One then obtains

$$\left| QF(p + \Phi_i) - QF(p + \Phi_{i-1}) \right| \leq \lambda_{n+1}^{-\beta} \left| A^\beta [F(p + \Phi_i) - F(p + \Phi_{i-1})] \right|$$
$$\leq \lambda_{n+1}^{-\beta} C_1 \nu^i C_5 = \nu^{i+1} C_5.$$

Also note that $|A^2 p| \leq \lambda_n |Ap|$ for $p \in P\mathcal{H}$. Since $\lambda_n \geq 1$ and $|Ap| \leq \rho$, one has

$$\left| A[PF(p+\Phi_i) - APp] \right| \leq \left| (AP)^\alpha \right| \left| A^\beta F(p+\Phi_i) \right| + \left| A^2 p \right| \leq \lambda_n^\alpha C_0 + \lambda_n \rho \leq \lambda_n (C_0 + \rho).$$

Turning to the first derivatives one finds that

$$AD\Phi_i(p)v = QDF(p + \Phi_{i-1})(P + D\Phi_{i-1}(p)P)v.$$

As a result, one has

$$|D\Phi_i(p)v| \leq \lambda_{n+1}^{-1} |AD\Phi_i(p)v| \leq \lambda_{n+1}^{-1}(\nu + \cdots + \nu^i)|Av| \leq 2\lambda_{n+1}^{-1} \nu |Av|,$$
$$|(AD\Phi_i(p) - AD\Phi_{i-1}(p))v| \leq C_7 \nu^i |Av|,$$

where $C_7 = 1 + 6C_2 C_1^{-1}$. As a result one has

$$|D\Phi_i[PF(p + \Phi_{i-1}) - APp]| \leq 2\lambda_{n+1}^{-1} \nu \lambda_n (C_0 + \rho) \leq 2\nu (C_0 + \rho),$$

and therefore

$$|E_i(p)| \leq \nu \left[\nu^i C_5 + 2(C_0 + \rho) \right] \leq \lambda_{n+1}^{-\beta} C_3,$$

where $C_3 = C_1 [C_5 + 2(C_0 + \rho)]$. This completes the proof of (5.8).

The derivative $DE_i(p)$ is given by

(5.10)
$$DE_i(p)v = D^2 \Phi_i [PF(p + \Phi_{i-1}) - APp]v$$
$$+ D\Phi_i \left[PDF(p + \Phi_i)(P + D\Phi_{i-1}P) - AP \right] v$$
$$+ QDF(p + \Phi_{i-1})(P + D\Phi_{i-1}P)v - QDF(p + \Phi_i)(P + D\Phi_i P)v$$

Now

$$AD^2 \Phi_i(p)vw = QD^2 F(p + \Phi_{i-1})(P + D\Phi_{i-1}(p)P)v(P + D\Phi_{i-1}(p)P)w$$
$$+ QDF(p + \Phi_{i-1})D^2 \Phi_{i-1} P)vw,$$

which implies that

$$|D^2\Phi_i(p)vw| \le \lambda_{n+1}^{-1}|AD^2\Phi_i(p)vw| \le \lambda_{n+1}^{-1}C_6(\nu+\cdots+\nu^i)|Av||Aw| \le 2C_6\lambda_{n+1}^{-1}\nu|Av||Aw|,$$

where $C_6 = 2C_2C_1^{-1}$. Consequently one has

$$|D^2\Phi_i[PF(p+\Phi_{i-1}) - APp]v| \le 2C_6\lambda_{n+1}^{-1}\nu\lambda_n(C_0+\rho)|Av| \le 2C_6\nu(C_0+\rho)|Av|.$$

Also one obtains

$$\begin{aligned}
|D\Phi_i PDF(p+\Phi_{i-1})(P+D\Phi_{i-1})v| &\le 2\lambda_{n+1}^{-1}\nu|APDF(p+\Phi_{i-1})(P+D\Phi_{i-1})v| \\
&\le 2\lambda_{n+1}^{-1}\nu\lambda_n^\alpha|A^\beta DF(p+\Phi_{i-1})(P+D\Phi_{i-1})v| \\
&\le 2\lambda_{n+1}^{-\beta}\nu C_1|A(P+D\Phi_{i-1})v| \\
&\le 2\nu^2(1+\nu+\cdots+\nu^i)|Av| \le 4\nu^2|Av|.
\end{aligned}$$

Next one has $|D\Phi_i APv| \le 2\lambda_{n+1}^{-1}\nu|AP||Av| \le 2\nu|Av|$. The last two terms in (5.10) are handled by using the following identity

(5.11)
$$\begin{aligned}
QDF(p+\Phi_{i-1})(P+D\Phi_{i-1})v &- QDF(p+\Phi_i)(P+D\Phi_i)v \\
&= [QDF(p+\Phi_{i-1}) - QDF(p+\Phi_i)](P+D\Phi_{i-1})v \\
&\quad + QDF(p+\Phi_i)[D\Phi_{i-1} - D\Phi_i]v
\end{aligned}$$

For the first term on the right-hand side of (5.11) one has

$$\begin{aligned}
|[QDF(p+\Phi_{i-1})-QDF(p+\Phi_i)](P+D\Phi_{i-1})v| & \\
&\hspace{-6cm}\le \lambda_{n+1}^{-\beta}|A^\beta[DF(p+\Phi_{i-1}) - DF(p+\Phi_i)](P+D\Phi_{i-1})v| \\
&\hspace{-6cm}\le \lambda_{n+1}^{-\beta}C_2|A(\Phi_{i-1} - \Phi_i)||A(P+D\Phi_{i-1})v| \\
&\hspace{-6cm}\le \lambda_{n+1}^{-\beta}C_2\nu^i C_5(1+\nu+\cdots+\nu^{i-1})|Av| \le \nu^{i+1}C_8|Av|,
\end{aligned}$$

where $C_8 = C_5C_2C_1^{-1}$. For the second term one obtains

$$|QDF(p+\Phi_i)[D\Phi_{i-1} - D\Phi_i]v| \le \lambda_{n+1}^{-\beta}C_1|A(D\Phi_{i-1} - D\Phi_i)v| \le C_7\nu^i|Av|.$$

Putting all these inequalities together one then finds that

$$|DE_i(p)v| \le \nu\left[2C_6(C_0+\rho) + 4\nu + 2 + \nu^i C_8 + \nu^{i-1}C_7\right]|Av| \le \lambda_{n+1}^{-\beta}C_4|Av|$$

where $C_4 \le C_1[2C_6(C_0+\rho) + 6 + C_8 + C_7]$. □

Since the constants C_3 and C_4 do not depend on i, we conclude that for the limiting functions $\Phi_\infty = \lim_{i\to\infty}\Phi_i$ and

$$E_\infty(p) \stackrel{\text{def}}{=} D\Phi_\infty[PF(p+\Phi_\infty) - APp]$$

also satisfy (5.8) and (5.9). In other words, $\mathfrak{M}_\infty = $ graph Φ_∞, where Φ_∞ is the solution of (5.5), is also an AIM with the same defect as the approximants, $\mathfrak{M}_i = $ graph Φ_i, $1 \le i$.

5.3 Comparisons. The classical Galerkin method as well as any of the non-linear Galerkin methods can be used to find AIMs of any defect $\eta > 0$. In order to do this one needs

$$|E_i(p)| \leq \eta, \qquad \text{and} \qquad |DE_i(p)| \leq \eta.$$

From (5.4), (5.8) and (5.9) we see that this will occur if one has

$$(5.12) \qquad\qquad \eta^{-1} C_9 \leq \lambda_{n+1}^{\beta}$$

for some constant C_9. Since $\beta > 0$ and $\lambda_n \to \infty$ as $n \to \infty$, we see that for any $\eta > 0$ (5.12) is valid for n sufficiently large.

Inequality (5.12) gives some information about the dimension of the AIM. For example, if the asymptotic relationship

$$(5.13) \qquad\qquad \lambda_n \sim C n^k, \qquad \text{as } n \to \infty$$

is satisfied for some $k > 0$, it follows from (5.12) that if n satisfies

$$\eta^{-1} C_{10} \leq n^{k\beta}$$

for some constant C_{10}, then the corresponding AIM has defect $\leq \eta$.

5.4 Other methods. There are other methods, variations of the methods described above, which have been used for the construction of AIMs. For example, there are methods which attempt to augment the nonlinear Galerkin methods by assigning a computable, but nonzero, value to q' and then solving

$$q' + AQq = QF(p+q)$$

for q. For example, let $\hat{\Phi}_0 = \Phi_0$ and $\hat{\Phi}_1 = \Phi_1$ be constructed as above. For the ith step we set

$$q'_{i-1} = D\Phi_{i-1}[PF(p + \Phi_{i-1}) - APp]$$

and then solve

$$AQ\hat{\Phi}_i = QF(p + \hat{\Phi}_{i-1}) - D\hat{\Phi}_{i-1}[PF(p + \hat{\Phi}_{i-1}) - APp]$$

for $\hat{\Phi}_i = \hat{\Phi}_i(p)$. This leads to a new family of approximations to the limiting inertial manifold, which is a solution of (2.9).

In addition, the manifolds $\hat{\mathfrak{M}}_i = \text{graph } \hat{\Phi}_i$ are AIMs for (2.1) because $\hat{\mathfrak{M}}_i$ is an inertial manifold for the perturbed equation

$$p' + APp = PF(p+q), \qquad q' + AQq = QF(p+q) + \hat{E}_i(p)$$

where

$$\hat{E}_i = D\hat{\Phi}_i[PF(p + \hat{\Phi}_i) - APp] + AQ\hat{\Phi}_i - QF(p + \hat{\Phi}_i).$$

One can show that if F is a C^2-function satisfying (5.7), then estimates similar to (5.8) and (5.9) are valid for $\hat{E}_i$.

There are also hybrid methods which combine various approximation methods. First there is the Euler-Galerkin method, which is described in Foias, Sell and Titi (1989), and which is implemented for the Kuramoto-Sivashinsky equation in Foias, Jolly, Kevrekidis, Sell, and Titi (1988) and Jolly, Kevrekidis, and Titi (1989). Secondly, there are the methods based on elliptic regularization, see Fabes, Luskin, and Sell (1988). Another method is to combine the classical Galerkin and the nonlinear Galerkin methods, see Marion and Temam (1989, 1990) and Temam (1989).

6. The Gamma method. Our next objective is to present a general method for the construction of an AIM of any defect η for the nonlinear evolutionary equation (2.1). We call this method the **Gamma Method**. As before, we assume that (2.1) satisfies the Standing Hypotheses.

Let P and Q be the orthogonal projections defined in Section 2, and consider the equivalent system

$$(6.1) \qquad p' + APp = PF(p+q), \qquad q' + AQq = QF(p+q).$$

Consider also the perturbed system

$$(6.2) \qquad p' + APp = PF(p+q), \qquad q' + \gamma AQq = QF(p+q),$$

where $\gamma > 1$. The term AQq appearing in (6.1) is a stable term, and the perturbation, which replaces AQq by γAQq, creates a system with a stronger stable term.

Let α be defined by $\alpha + \beta = 1$, where β is given by the Standing Hypotheses. Note that one has $0 \le \alpha < 1$. By using a variation of the argument in Foias, Sell, and Temam (1988), we conclude that there are constants K_0 and K_1 such that if the eigenvalues λ_n and λ_{n+1} satisfy $\lambda_n \ge K_0$, together with the gap condition

$$(6.3) \qquad \gamma\lambda_{n+1} - \lambda_n \ge K_1 \left(\gamma^\alpha \lambda_{n+1}^\alpha + \lambda_n^\alpha\right),$$

then (6.2) has an inertial manifold of the form

$$\mathfrak{M}_\gamma = \text{graph } \Phi_\gamma,$$

where Φ_γ is a smooth function from $P\mathcal{H}$ to $Q\mathcal{H} \cap \mathcal{D}(A)$. Furthermore Φ_γ satisfies the following properties

$$(6.4) \qquad |A\Phi_\gamma(p)| \le \hat{C}_1 \lambda_{n+1}^{-\beta}, \qquad p \in P\mathcal{H}.$$

for some constant $\hat{C}_1$, and

$$(6.5) \qquad |AD_p\Phi_\gamma(p)h| \le |Ah|, \qquad p \in P\mathcal{H}, \quad h \in \mathcal{D}(A).$$

Now define η by

$$\eta = (\gamma - 1)\max\left(\hat{C}_1 \lambda_{n+1}^{-\beta}, 1\right),$$

and set $\Phi_\eta = \Phi_\gamma$.

We claim that $\mathfrak{M}_\eta = \text{graph } \Phi_\eta$ is an AIM for (2.1), or (6.1), of defect η. Indeed, consider the equation

$$(6.6) \qquad p' + APp = PF(p+q), \qquad q' + AQq = QF(p+q) + E(p),$$

where $E(p) \stackrel{\text{def}}{=} -(\gamma - 1)AQ\Phi_\eta(p)$. It follows from (6.4) and (6.5) that the perturbation term $E(p)$ satisfies

(6.7)
$$\sup_{p \in P\mathcal{H}} |E(p)| \leq \eta, \qquad \sup_{p \in P\mathcal{H}} |D_p E(p)| \leq \eta.$$

Furthermore, if one sets $q = \Phi_\eta(p)$, then $q' = D\Phi_\eta p'$ and

(6.8)
$$\begin{aligned} D\Phi_\eta (PF(p + \Phi_\eta) - APp) &= QF(p + \Phi_\eta) - \gamma AQ\Phi_\eta \\ &= QF(p + \Phi_\eta) - AQ\Phi_\eta + E(p), \end{aligned}$$

i.e., $\mathfrak{M}_\eta = $ graph Φ_η is invariant for (6.6), as well as (6.2).

It remains to prove that the AIM $\mathfrak{M}_\eta = $ graph Φ_η is exponentially attracting when $\lambda_{n+1}^\beta > C_1$. The proof is a minor adaptation of the argument in Section 3. The main change is in the handling of the change of variables $q = \Phi_\eta + r$ in (6.6), where q is a solution of

$$q' + AQq = QF(p + \Phi_\eta + r) + E(p),$$

and

$$\Phi_\eta' = D\Phi_\eta \frac{\partial}{\partial t} p = D\Phi_\eta (PF(p + \Phi_\eta) - APp).$$

It follows from (6.8) that

$$\Phi_\eta' + AQ\Phi_\eta = QF(p + \Phi_\eta) + E(p).$$

As a result, we see that in the (p, r)-variables, the system (6.6) now takes the form

$$p' + APp = PF(p + \Phi_\eta + r), \qquad r' + AQr = QF(p + \Phi_\eta + r) - QF(p + \Phi_\eta).$$

The remainder of the argument is now identical to that given in Section 3.

7. Applications. The first issue which arises, when one wants to apply any of the AIM methods, including the Gamma Method, to nonlinear evolutionary equations like the Navier-Stokes equations, is the requirement of bounded support in (2.2). This is not a serious issue. Let $\mathfrak{A}$ be the given attractor for (2.1) and let $\mathcal{O}$ be the given positively invariant neighborhood of $\mathfrak{A}$ in $\mathcal{D}(A)$, where $\mathcal{O}$ is bounded in $\mathcal{D}(A)$. This means that there is a $\rho > 0$ so that

$$\mathfrak{A} \subset \mathcal{O} \subset \{u \in \mathcal{D}(A) : |Au| \leq \frac{\rho}{2}\}.$$

One can then study a modified equation, which satisfies (2.2) as well as agreeing with the original equation in $\{u \in \mathcal{D}(A) : |Au| \leq \frac{\rho}{2}\}$. Furthermore the given attractor $\mathfrak{A}$ is an attractor for the modified equation, and $\mathcal{O}$ is a positively invariant set which lies within the basin of attraction for $\mathfrak{A}$ in the modified equation.

The second issue which arises is the the spectral gap condition (7.2). This is the main issue we address in the following examples.

7.1. Navier-Stokes equations. The Navier-Stokes equations on a bounded (minimally smooth) region $\Omega \subset R^k$, where $k = 2, 3$, are given by

$$(7.1) \qquad u_t - \nu\Delta u + (u \cdot \nabla)u + \nabla p = f, \qquad \nabla \cdot u = 0$$

in Ω and $u = 0$ on $\partial\Omega$, see Constantin and Foias (1988), Ladyzhenskaya (1972), and Temam (1977, 1983, 1988a). We also consider (7.1) with periodic boundary conditions on the rectangle in R^2 or a parallelepiped in R^3. We use the approach employed in the references cited above and project (7.1) into the space of divergent-free vector fields on Ω. In this way (7.1) becomes

$$(7.2) \qquad u' + \nu Au + B(u, u) = f$$

on the space $\mathcal{H}$ of divergent-free vector fields in $L^2(\Omega)$. (In the periodic case, we also require that the forcing term f and the functions in $\mathcal{H}$ have mean-value 0.) In (7.2) one has $A = -P_0\Delta$, where P_0 is the orthogonal projection onto $\mathcal{H}$, and $B(u, u)$ is a bilinear form that satisfies

$$B : \mathcal{D}(A) \times \mathcal{D}(A) \to \mathcal{D}(A^{1/2}).$$

Furthermore, there is a constant C_2 such that

$$(7.3) \qquad |A^{1/2}B(u, v)| \le C_2|Au||Av|, \qquad \text{for all } u, v \in \mathcal{D}(A).$$

We assume that the forcing function f is time-independent and $f \in \mathcal{D}(A^{1/2})$. This means that, with the exception of the local dissipative property, the equation (7.2) satisfies the Standing Hypotheses with $\beta = \frac{1}{2}$.

For the 2D problem, equation (7.2) has a global attractor $\mathfrak{A} \subset \mathcal{D}(A)$ and the basin of attraction for $\mathfrak{A}$ is the entire space $\mathcal{H}$, see the references cited above, as well as Foias, Sell, and Temam (1988). Furthermore, for any minimally smooth region $\Omega \subset R^k$, for any $k \ge 1$, there is a constant D_k such that the eigenvalues λ_n of $-\Delta$ have the asymptotic representation

$$(7.4) \qquad \lambda_n = D_k n^{\frac{2}{k}} + o(n^{\frac{1}{k}}), \qquad \text{as } n \to \infty,$$

see Edmund and Evans (1987) and Sell (1989). The eigenvalues for the Stokes operator A in (7.2) have the same asymptotic expansion for $k = 2, 3$. The two terms in the gap condition (7.2) for the 2D Navier-Stokes equations then become

$$\gamma\lambda_{n+1} - \lambda_n = \nu D_2(\gamma - 1)n + o(n^{\frac{1}{2}})$$

and

$$K_1(\gamma^{\frac{1}{2}}\lambda_{n+1}^{\frac{1}{2}} + \lambda_n^{\frac{1}{2}}) = o(n^{\frac{1}{2}}).$$

Consequently for each $\gamma > 1$, there exists infinitely many n such that (7.2) is satisfied. Hence the Navier-Stokes equations in 2D have AIMs of any defect $\eta > 0$.

For the 3D Navier-Stokes equations, there is a fundamental issue concerning the global regularity of solutions. Recently it was shown that for certain thin 3D domains, there is a local attractor $\mathfrak{A}$ with a *large* basin of attraction in $\mathcal{D}(A)$, see Raugel and Sell (1990). For this 3D problem the gap condition becomes

$$\gamma\lambda_{n+1} - \lambda_n = \nu D_3(\gamma - 1)n^{\frac{2}{3}} + o(n^{\frac{1}{3}})$$

and

$$K_1(\gamma^{\frac{1}{2}}\lambda_{n+1}^{\frac{1}{2}} + \lambda_n^{\frac{1}{2}}) = o(n^{\frac{1}{3}}).$$

Consequently for each $\gamma > 1$, there exists infinitely many n such that (7.2) is satisfied. Hence the Navier-Stokes equations for the thin domains in 3D have AIMs of any odefect $\eta > 0$.

7.2. Reaction diffusion equations. We are interested in systems of reaction diffusion equations of the form

$$(7.5) \qquad u_t = D\triangle u + f(x, u)$$

on a minimally smooth bounded region $\Omega \subset R^k$, $k \geq 1$, where $u \in R^m$ and D is an $m \times m$ diagonal matrix with positive entries on the diagonal. We consider (7.5) with either Dirichlet, Neumann or periodic boundary conditions We assume that (7.5) admits an invariant region in R^m and that $f : \Omega \times R^m \to R^m$ is a sufficiently smooth function, see Temam (1988a) and Marion (1989). It follows then that (7.5) generates a nonlinear evolutionary equation of the form (2.1) where the Standing Hypotheses are satisfied with $\beta = 1$. Let $\mathfrak{A}$ be the attractor in $\mathcal{D}(A)$ associated with the invariant region. Since $\alpha = 0$, the gap condition (7.2) reduces to

$$\gamma\lambda_{n+1} - \lambda_n \geq K_2.$$

From (7.4) we obtain

$$\gamma\lambda_{n+1} - \lambda_n = \hat{D}_k n^{\frac{2}{k}} + o(n^{\frac{1}{k}}).$$

We see then that for each K_2 there are infinitely values of n that satisfy the spectral gap condition (7.2).

This application to reaction diffusion equations is especially interesting because it is known that there does exist a reaction diffusion equation on the 4-dimensional set $[0, 2\pi]^4 \subset R^4$ which has a finite dimensional global attractor and which does not have a normally hyperbolic inertial manifold, see Mallet-Paret, Sell, and Shao (1991).

7.3 Advantage of the Gamma method. In order to compare the linear and nonlinear Galerkin methods with the Gamma Method, it is convenient to look at the gap condition (6.3) under the assumption that (5.13) holds. In this case, the left side of (6.3) becomes

$$\gamma\lambda_{n+1} - \lambda_n \sim C(\gamma - 1)n^k,$$

while the right side is

$$K_1(\gamma^\alpha \lambda_{n+1}^\alpha + \lambda_n^\alpha) \sim C_{11} n^{k\alpha}.$$

If we set $\eta = (\gamma - 1)$, then the gap condition (6.3) becomes

$$n^{k\beta} = n^{k-k\alpha} \geq C_{12}\eta^{-1},$$

for some constant C_{12}. Because of (6.7) we see that the Gamma Method, like the Galerkin methods, has the property that if n satisfies

$$\eta^{-1} C_{13} \leq n^{k\beta}$$

for some constant C_{13}, then the corresponding AIM has defect $\leq \eta$.

We see then that all the methods described above can be used to construct AIMs for (2.1) of any defect $\eta > 0$. If the eigenvalues of A satisfy (5.13), then all these AIMs have roughly the same dimensions. In order to further distinguish between these methods, one needs to use other criteria. The classical Galerkin method is certainly the simplest. The advantages of the nonlinear Galerkin methods are discussed in Foias, Manley, and Temam (1988), Marion (1989a, 1989b), Marion and Temam (1989 1990), Temam (1989), and Titi (1990). One point worth emphasizing here is that the AIM given by $\mathfrak{M}_\infty = $ graph Φ_∞, where Φ_∞ is the solution of (5.5), has the property that every stationary solution of (2.1) lies on $\mathfrak{M}_\infty$.

Nevertheless, we feel that the Gamma Method has a major advantage over all the Galerkin type methods described above. The principle advantage of the Gamma Method can be seen in the case where (2.1) has an inertial manifold and the gap condition (6.3) holds with $\gamma = 1$, i.e., one has

(7.6) $$\lambda_{n+1} - \lambda_n \geq K_1(\lambda_{n+1}^\alpha + \lambda_n^\alpha)$$

for some $n \geq 1$. This is the situation one encounters with the Kuramoto-Sivashinsky equation, see Foias, Nicolaenko, Sell and Temam (1988), and with certain reaction diffusion equations in low space dimension, see Foias, Sell and Temam (1988) and Mallet-Paret and Sell (1988). For $\gamma > 1$, the gap condition (6.3) becomes

(7.7) $$(\gamma - 1)\lambda_{n+1} + (\lambda_{n+1} - \lambda_n) \geq K_1(\gamma^\alpha \lambda_{n+1}^\alpha + \lambda_n^\alpha).$$

One can expect therefore, that (7.7) will hold for smaller values of n than in (7.6). If this happens, then one can find a good AIM of lower dimension than the inertial manifold!

8. An illustrative example. We consider a system of two reaction diffusion equations on the interval $\Omega = [0, 2\pi]$ of the form

(8.1)
$$u_t = u_{xx} + \rho^2(u - a^2 u^3) + f(u, v)$$
$$v_t = v_{xx} + \rho^2(v - a^2 v^3) + g(u, v)$$

with Dirichlet boundary conditions. We assume that ρ is a noninteger satisfying $\rho > 1$ and $0 < a \leq 1$. Furthermore we asume that f and g are C^4-functions that satisfy

$$|f(u,v)| \leq 1, \qquad |g(u,v)| \leq 1$$

for all $u, v \in \mathbf{R}$. Furthermore we ask that all four derivatives of f and g be bounded by ρ. An elemenary argument based on the maximum principle shows that

$$\left\{ (u,v) \in R^2 : |u|, |v| \leq \frac{a+1}{a} \right\}$$

is an invariant region for (8.1).

The system (8.1) converts to a nonlinear evolutionary equation

(8.2) $$U' + AU = F(U),$$

where

$$U = \begin{pmatrix} u \\ v \end{pmatrix}, \qquad A = -\Delta \begin{pmatrix} 1 & 0 \\ 0 & 1 \end{pmatrix},$$

$$F(U)(x) = \begin{pmatrix} \rho^2(u(x) - a^2 u^3(x)) + f(u(x), v(x)) \\ \rho^2(v(x) - a^2 v^3(x)) + g(u(x), v(x)) \end{pmatrix}.$$

The eigenvalues of A are $\lambda_n = n^2$, $n = 1, 2, \ldots$, and each of these eigenvalues has multiplicity 2. The global attractor for (8.1), or (8.2), lies in a bounded set in $\mathcal{D}(A) = \left(H^2(\Omega) \cap H_0^1(\Omega) \right)^2$. Also (8.2) satisfies the Standing Hypotheses with $\beta = 1$.

One can show that the gap condition (5.3) for (8.2) assumes the form

(8.3) $$\gamma \lambda_{n+1} - \lambda_n \geq K_1 \rho,$$

where K_1 is a fixed universal constant. For $\gamma = 1$, the gap condition (8.3) becomes

$$2n + 1 \geq K_1 \rho.$$

Since the eigenvalues have multiplicity 2, it follows that the inertial manifold $\mathfrak{M}$ for (8.2) has dimension $2n$, or $\dim \mathfrak{M} = K_1 \rho - 1$. Now let $\gamma > 1$ and set $\eta = (\gamma - 1)$. Then the gap condition (8.3) becomes

(8.4) $$\eta m^2 + 2m = K_1 \rho,$$

where $m = n + 1$.

REFERENCES

P. CONSTANTIN AND C. FOIAS (1988), *Navier-Stokes Equations*, Univ. Chicago Press, Chicago.

D. E. EDMUNDS AND W. D. EVANS (1987), *Spectral Theory and Differential Operators*, Oxford Math. Monographs.

E. FABES, M. LUSKIN AND G. R. SELL (1991), *Construction of inertial manifolds by elliptic regularization*, IMA Preprint No. 459, J. Differential Equations, (to appear).

N. FENICHEL (1971), *Persistence and smoothness of invariant manifolds for flows*, Indiana Univ. Math. J., 21, pp. 193-226.

C. FOIAS, M. S. JOLLY, I. G. KEVREKIDIS, G. R. SELL, E. S. TITI (1988), *On the computation of inertial manifolds*, Physics Letters A, 131, pp. 433-436.

C. FOIAS, O. MANLEY, AND R. TEMAM (1988), *Modelling of the interaction of small and large eddies in two dimensional turbulent flows*, Math. Modelling Numerical Anal., 22, pp. 93-118.

C. FOIAS, B. NICOLAENKO, G. R. SELL, AND R. TEMAM (1988), *Inertial manifolds for the Kuramoto Sivashinsky equation and an estimate of their lowest dimensions*, J. Math. Pures Appl., 67, pp. 197-226.

C. FOIAS, G. R. SELL AND R. TEMAM (1988), *Inertial manifolds for nonlinear evolutionary equations*, J. Differential Equations, 73, pp. 309-353.

C. FOIAS, G. R. SELL AND E. S. TITI (1989), *Exponential tracking and approximation of inertial manifolds for dissipative equations*, J. Dynamics and Differential Equations, 1, pp. 199-244.

C. FOIAS AND R. TEMAM (1988), *The algebraic approximation of attractors: The finite dimensional case*, Physica D, 32, pp. 163-182.

J. K. HALE (1988), *Asymptotic Behavior of Dissipative Systems*, Math. Surveys and Monographs, Vol. 25, Amer. Math. Soc., Providence, R. I..

D. B. HENRY (1981), *Geometric Theory of Semilinear Parabolic Equations*, Lecture Notes in Mathematics, No. 840, Springer Verlag, New York.

M. S. JOLLY, I. G. KEVREKIDIS, AND E. S. TITI (1989), *Approximate inertial manifolds for the Kuramoto-Sivashinsky equation: Analysis and computations*, MSI Technical Report 89-52, Cornell Univ.

O. A. LADYZHENSKAYA (1972), *On the dynamical system generated by the Navier-Stokes equations*, English translation, J. Soviet Math., 3, pp. 458-479.

M. LUSKIN AND G. R. SELL (1988), *Approximation theories for inertial manifolds*, Proc. Luminy Conference on Dynamical Systems, Math. Modelling Numerical Anal., 23.

J. MALLET-PARET AND G. R. SELL (1988), *Inertial manifolds for reaction diffusion equations in higher space dimensions*, J. Amer. Math. Soc., 1, pp. 805-866.

J. MALLET-PARET, G. R. SELL, AND Z. SHAO (1991), *Counterexamples to the existence of inertial manifolds*, In preparation.

M. MARION (1989A), *Approximate inertial manifolds for reaction diffusion equations in high space dimension*, J. Dynamics and Differential Equations, 1, pp. 245-267.

M. MARION (1989B), *Approximate inertial manifolds for the pattern formation in the Cahn-Hilliard equation*, Proc. Luminy Conference on Dynamical Systems, Math. Modelling Numerical Anal., 23, pp. 463-488.

M. MARION AND R. TEMAM (1989), *Nonlinear Galerkin methods*, SIAM J. Numerical Anal., 26, pp. 1139-1157.

M. MARION AND R. TEMAM (1990), *Nonlinear Galerkin methods: The finite elements case*, Numerische Math., 57, pp. 205-226.

V. A. PLISS (1977), *Integral Sets of Periodic Systems of Differential Equations*, Russian, Izdat. Nauka, Moscow.

V. A. PLISS AND G. R. SELL (1990), *Perturbations of attractors of differential equations*, IMA Preprint No. 680, J. Differential Equations(to appear).

G. RAUGEL AND G. R. SELL (1990), *Navier-Stokes equations in thin 3D domains: Global regularity of solutions I*, IMA Preprint No. 662.

R. J. SACKER (1969), *A perturbation theorem for invariant manifolds and Hölder continuity*, J. Math. Mech., 18, pp. 705-762.

G. R. SELL (1989), *Hausdorff and Lyapunov dimensions for gradient systems*, in *The Connection between Infinite Dimensional and Finite Dimensional Dynamical Systems*, Contemporary Mathematics, Vol. 99, pp. 85-92.

R. TEMAM (1977), *Navier-Stokes Equations*, North-Holland, Amsterdam.

R. TEMAM (1983), *Navier-Stokes Equations and Nonlinear Functional Analysis*, CBMS Regional Conference Series, No. 41, SIAM, Philadelphia.

R. Temam (1988), *Infinite Dimensional Dynamical Systems in Mechanics and Physics*, Springer Verlag, New York.

R. Temam (1989), *Attractors for the Navier-Stokes equations: Localization and approximation*, J. Faculty Sci. Tokyo, Sec 1A, 36, pp. 629-647.

E. S. Titi (1990), *On approximate inertial manifolds to the 2D Navier-Stokes equations*, J. Math. Anal. Appl., 149, pp. 540-557.

SOME RECENT RESULTS ON
INFINITE DIMENSIONAL DYNAMICAL SYSTEMS

ROGER TEMAM*

Abstract. The object of this article is to survey some recent developments in the theory of infinite dimensional dynamical systems. We shall successively consider the derivation of optimal bounds for the dimension of the attractor for the Navier-Stokes equations in space dimension three; new developments in the theory of inertial manifolds and their connection to the concept of slow manifolds broadly used in meteorology; and the approximation of inertial manifolds using finite differences in connection with multigrid methods and wavelets.

Introduction. In this article we survey some recent developments in the theory of infinite dimensional dynamical systems.

The first result that we present (Sec. 1) concerns the derivations of optimal bounds for the attractor of the three-dimensional Navier-Stokes equations. The second result presented in Sec. 2 is an existence result of inertial manifolds, generalizing that of Foias-Sell-Temam (1985, 1988) to the case where the linear operator is not self-adjoint. The principal application and motivation of this result lies in the concept of slow manifold. Slow manifolds are broadly used in meteorology for short term weather forecast. By the results presented in Sec. 2, slow manifolds appear simply as a particular case of inertial manifolds. From the mathematical viewpoint we meet here challenging problems involving spectral theory of linear operators and inertial manifold theory.

In Sec. 3 we give some indications on current research on the approximation of inertial manifolds using finite differences. We recall the concept of Incremental Unknowns (IU) introduced in Temam (1990a) and how the IU method can be used to construct approximate inertial manifolds. The IU method which naturally relates to the multigrid method also relates to wavelets in a way which will be described and investigated in a forthcoming article.

Content

1. Optimal bounds of the dimension of the Navier-Stokes attractor in $\mathbf{R}^3$

2. Inertial manifolds and Slow manifolds

3. Incremental Unknowns.

1. Optimal bounds of the dimension of the Navier-Stokes attractor in. $\mathbf{R}^3$. It has been proved by Foias-Temam (1979) that attractors for the $2D$ Navier-Stokes equations have finite Hausdorff (and fractal) dimension. The same is true in space dimension three for invariant sets and attractors that are bounded in the enstrophy norm. Furthermore physically relevant estimates on the dimension

---

*Laboratoire d'Analyse Numérique, Université Paris-Sud, Bât. 425, 91405 Orsay, France.

were derived in Constantin-Foias-Manley-Temam (1985) for space dimension three and in Constantin-Foias-Temam (1988) for space dimension two.

In space dimension three it was shown that the dimension of an attractor is of order

$$(1.1) \qquad \dim \mathcal{A} \sim c \left(\frac{L_0}{L_d} \right)^3$$

where c is an absolute constant, L_0 a macroscopic length of the flow and L_d the Kolmogorov dissipation length. Actually this result produced the first connection between two approaches to turbulence in fluids which seemed unrelated, namely the conventional approach and the dynamical systems approach.

Result (1.1) shows that the estimates on attractor dimensions in Constantin-Foias-Manley-Temam (1985) agree fully with physical results and are, in this sense, optimal. Here, following Ghidaglia-Temam (1991) we want to show that these estimates are also optimal in a more mathematical sense, by presenting examples of three dimensional turbulent flows where the upper bound and the lower bound of the dimension of $\mathcal{A}$ agree, i.e. are of the same order. Of course since the mathematical theory of the $3D$ Navier-Stokes equations is not yet complete, the proof of this result is not complete ; however it relies as we shall see below on some physically reasonable hypotheses.

Let us now describe the study contained in Ghidaglia-Temam (1991). We consider a space periodic flow in $\mathbf{R}^3$ of Poiseuille type. Namely the periods are $2\pi/\alpha$, 2π, $2\pi/\beta$ in directions x_1, x_2, x_3, with $0 < \beta \leq \alpha \leq 1$, and α is intended to converge to zero. Hence the parallelepiped $\Omega = (0, 2\pi/\alpha) \times (0, 2\pi) \times (0, 2\pi/\beta)$ is elongated in two directions.

The flow is driven by volume forces

$$f(x_1, x_2, x_3) = (g(x_2), 0, 0)$$

where g is a smooth function satisfying

$$(1.2) \qquad \int_0^{2\pi} g(x_2) dx_2 = 0.$$

It is easy to see that the Navier-Stokes equations possess a unique shear flow stationary solution u_S such that $u_S(x_1, x_2, x_3) = (U(x_2), 0, 0)$,

$$(1.3) \qquad \begin{aligned} -\nu \frac{d^2 U}{dx_2^2}(x_2) &= g(x_2), \\ \int_0^{2\pi} U(x_2) dx_2 &= 0. \end{aligned}$$

The aim is to study the linear stability of the stationary solution u_S and by deriving a lower bound on the dimension of its unstable manifold $\mathcal{M}_u$ to derive a lower bound on the dimension of invariant manifolds (or attractor) containing $\mathcal{M}_u$; and to compare this lower bound to the available upper bounds.

The study generalizes a similar study due to Babin-Vishik (1983) in space dimension two. The result that we proved is the following

(1.4)

> *There exist two positive constants a_0 and d_0*
> *depending only on U/ν such that if*
>
> $$0 < \beta \leq \xi \leq 1, \quad \alpha^2 + \beta^2 \leq q_0^2$$
>
> *and*
>
> $$\nu^2 < \frac{1}{8\pi} \int_0^{2\pi} [\theta(y)]^2 dy$$
>
> *then*
>
> $$\dim \mathcal{A}_0 \geq \dim \mathcal{M}_u \geq \frac{d_0}{\alpha\beta}.$$

Here $\mathcal{A}_0$ is any invariant set or attractor containing the invariant manifold $\mathcal{M}_u$ of u_S; and θ is the $2\pi-$periodic primitive function of U such that

(1.5)
$$\int_0^{2\pi} \theta(y)dy = 0.$$

The proof of (1.4) relies on the study of the solutions of an Orr-Sommerfeld equation depending on a parameter.

Then we derive an upper bound on the dimension of the attractor and compare it to the lower bound. For that purpose we make the following hypotheses :

(i) <u>Nonoccurence of singularities</u>

The enstrophy norm (i.e. the L^2-norm of vector $\mathrm{curl}\ u$) remains bounded for all time on the attractor or invariant set $\mathcal{A}_0$ under consideration :

(1.6)
$$\operatorname*{Sup}_{t \in \mathbf{R}} \int_\Omega |\mathrm{curl}\ u(x,t)|^2 dx < \infty$$

for all orbits lying in $\mathcal{A}_0$ ($\mathcal{A}_0 \supset \mathcal{M}_u$, possibly $\mathcal{A}_0 = \mathcal{M}_u$).

(ii) <u>Boundedness of a Reynolds number</u>

It is assumed that the following Reynolds number based on the maximum of the magnitude of the velocity vector remains bounded independently of α, for all trajectories $u(\cdot)$ lying on $\mathcal{M}_u$ (or $\mathcal{A}_0$) :

(1.7)
$$Re = \frac{1}{\nu} \operatorname*{Sup}_u \limsup_{t \to \infty} \left(\frac{1}{t} \int_0^t \operatorname*{Sup}_{x \in \Omega} |u(x,s)|^2 dx \right)^{1/2}.$$

It was then shown that there exists an absolute constant c such that

(1.8)
$$\dim \mathcal{A}_0 \leq c \frac{1 + Re^3}{\alpha\beta}.$$

Hence if we are interested in the passage to the limit $\alpha, \beta \to 0$, with ν and f fixed we conclude that the upper and lower bound (1.8), (1.4) on the dimension of $\mathcal{A}_0$ agree

$$(1.9) \qquad \qquad \operatorname{dim}\mathcal{A}_0 \simeq \frac{K}{\alpha\beta}$$

Note also that this dimension is proportional to the volume of Ω and this should be compared to the well known and yet unproved conjecture of Pomeau and Manneville (1979) concerning the dimension of the attractor for the Kuramoto-Sivashinsky equation (i.e. that it is proportional to the length of the interval).

The proof of (1.8) relies on the methods of Constantin-Foias-Temam (1985) and on delicate inequalities of Lieb-Thirring type and inequalities concerning the eigenvalues of the Laplace operator in (thin) elongated domains.

Already in space dimension two, the study of Ghidaglia-Temam (1991) fills the small gap[1] remaining between the lower bound on the dimension of the global attractor for the Navier-Stokes equations due to Babin-Vishik (1983) and the upper bounds in Constantin-Foias-Temam (1988) and Temam (1985). By extending the proof of this last reference to the flow in elongated domains $\Omega = (0, 2\pi/\alpha) \times (0, 2\pi)$, it is shown that

$$(1.10) \qquad \qquad \operatorname{dim} \mathcal{A} \leq \frac{c}{\alpha}$$

in full agreement with the lower bound $\dim \mathcal{A} = c'/\alpha$ in Babin-Vishik's article. For other results on the dimension of attractors for the $2D$ Navier-Stokes equations see Doering-Gibbon (1991).

Finally let us make a remark concerning the physical validity of hypotheses (i) and (ii). Hypothesis (i) is equivalent to the nonoccurence of singularities in the flow (see e.g. Temam (1977)). Hypothesis (ii) is reasonable since it can be rigorously proved that, due to the special form of the forces f, a related Reynolds number remains bounded independently of α. Indeed, let

$$Re' = \frac{1}{\nu} \operatorname*{Sup}_{u} \limsup_{t\to\infty} \left(\frac{1}{t} \int_0^t \frac{1}{|\Omega|} \int_\Omega |\nabla u(x,s)|^2 dx ds \right)^{1/2}.$$

The energy conservation equation for an orbit $u(\cdot)$ gives

$$\frac{1}{2}\frac{d}{dt} \int_\Omega |u|^2 dx + \nu \int_\Omega |\nabla u|^2 dx = \int_\Omega f u\, dx = \int_\Omega g(x_2) u_1(x) dx$$
$$= -\int_\Omega h(x_2) \frac{\partial u_1}{\partial x_2}(x) dx,$$

---

[1] The statement in Constantin-Foias-Temam (1988) (Remark 3.1(i)) saying that the upper bound in that reference agrees with the lower bound of Babin-Vishik up to a logarithmic correction is not correct. Indeed in the first reference the shape of Ω is fixed while the volume density of forces and/or ν^{-1} become large ; instead in the second reference the volume density of forces and the kinematic viscosity are fixed while the domain becomes elongated.

where

$$h(x_2) = \int_0^{x_2} g(s)ds.$$

The right-hand side of this equation is majorized thanks to the Cauchy-Schwarz inequality by

$$\left(\int_\Omega |h|^2 dx\right)^{1/2}\left(\int_\Omega \left|\frac{\partial u_1}{\partial x_2}\right|^2 dx\right)^{1/2}$$

$$\leq \frac{\nu}{2}\int_\Omega |\nabla u|^2 dx + \frac{1}{2\nu}\int_\Omega |h(x_2)|^2\, dx.$$

It follows readily that

$$\limsup_{t\to\infty}\frac{1}{t}\int_0^t \frac{1}{|\Omega|}\int_\Omega |\nabla u|^2 dx\,ds$$

$$\leq \frac{1}{\nu^2|\Omega|}\int_\Omega |h(x_2)|^2\, dx \leq \frac{1}{2\pi\nu^2}\int_0^{2\pi} |h(x_2)|^2\, dx_2.$$

Hence

$$Re' \leq \frac{1}{\nu^2}\left(\frac{1}{2\pi}\int_0^{2\pi}|h(x_2)|^2\, dx_2\right)^{1/2}.$$

2. Inertial manifolds and slow manifolds. Most articles deriving existence results for inertial manifolds are related to evolution equations of the form

$$(2.1) \qquad \begin{cases} \dfrac{du}{dt} + Au + R(u) = 0, \\ u(0) = u_0, \end{cases}$$

where the main linear operator A is self-adjoint positive unbounded in a Hilbert space H and has a compact inverse A^{-1} (see however Chow-Lu (1988) for a different setting).

There are however motivations for considering more general equations

$$(2.2) \qquad \begin{cases} \dfrac{dv}{dt} + \mathcal{A}v + \mathcal{R}(v) = 0 \\ v(0) = v_0 \end{cases}$$

where the operator $\mathcal{A}$ is a suitable compact perturbation of an operator like A. One of the motivations is the need to construct more general types of inertial manifolds, as explained below. Another motivation is to make the connection between Inertial Manifolds and the important concept, broadly used in meteorology, of Slow Manifolds (see e.g. Temam (1990b)). The following results are excerpts from Debussche-Temam (1991).

Since A^{-1} is compact and self-adjoint, there exists an orthonormal basis of H consisting of eigenvectors $\{w_j\}_{j\in\mathbf{N}}$ of A :

$$(2.3) \qquad \begin{cases} Aw_j = \lambda_j w_j, \quad j \in \mathbf{N} \\ 0 < \lambda_1 \leq \lambda_2 \leq ..., \quad \lambda_j \to \infty \text{ as } j \to \infty. \end{cases}$$

The operator R is assumed to be a $\mathcal{C}^1$ map from the space $D(A^\alpha)$ (domain of A^α in H) into H, for some α, $0 \le \alpha < 1$. Under suitable hypotheses (2.1) defines a continuous semigroup $\{S(t)\}_{t\ge 0}$ in $D(A^\alpha)$ that possesses an absorbing set B in $D(A^\alpha)$. Most inertial manifolds for equation (2.1) have been obtained as the graph of a function Φ :

$$\Phi \; : \; P_m H \to Q_m H,$$

where, for some integer m, $P_m H$ is the space spanned by $w_1, ..., w_m$, P_m the orthogonal projector onto this space and $Q_m = I - P_m$. (see e.g. Foias-Sell-Temam (1985, 88), Constantin-Foias-Nicolaenko-Temam (1988, 89), Mallet-Paret-Sell (1988)). We shall show how, by considering equations like (2.2) one can obtain other families of inertial manifolds for both equations, (2.2) and (2.1).

For equation (2.2) we assume that $\mathcal{A}$ is a lower order perturbation of A, i.e. $\mathcal{A} = A + b$ with b linear, $D(A^\alpha) \subset D(b)$ and $bA^{-\alpha}$ bounded, for the same α as before ; the operator $\mathcal{R}$ satisfies the same properties as R.

The spectrum of $\mathcal{A}$ consists of eigenvalues of finite multiplicity. Moreover if we denote by $\{\nu_j\}_{j\in\mathbb{N}}$ these eigenvalues ordered so that the sequence $\{Re(\nu_j)\}_{j\in\mathbb{N}}$ is nondecreasing, then it can be shown using results of Gohberg-Krein (1969) and hypothesis (2.7) below, that

$$Re(\nu_j) \sim \lambda_j, \quad \text{as } j \to \infty.$$

Let also $\{v_j\}_{j\in\mathbb{N}}$ be a system of root vectors associated to the v'_j s:

$$\forall j \in \mathbb{N}, \;\; \exists k_j \ge 1, \;\; (\mathcal{A} - \nu_j I)^{k_j} v_j = 0,$$

and let $\mathcal{P}_m$ denote the projector into the space spanned by $v_1, ..., v_n$ parallel to $v_{n+1}, ..., v_\ell, .$

Inertial manifolds for (2.2) are derived as graphs of functions Φ :

$$\Phi \; : \; \mathcal{P}_m H \to \mathcal{Q}_m H,$$

for some appropriate m, with $\mathcal{Q}_m = I - \mathcal{P}_m$. As we shall see this provides also inertial manifolds for equation (2.1) as well.

Indeed the following case is of particular interest. Let $\bar{u}$ be a stationary solution of (2.1)

$$(2.4) \qquad\qquad A\bar{u} + R(\bar{u}) = 0$$

and let $v = u - \bar{u}$. Then v is solution of an equation like (2.2) where

$$(2.5) \qquad\qquad \begin{aligned} \mathcal{A}v &= Av + DR(\bar{u}) \cdot v \\ \mathcal{R}(v) &= R(\bar{u} + v) - R(\bar{u}) - DR(\bar{u}) \cdot v \end{aligned}$$

where $DR(\bar{u})$ is the Fréchet differential of R at $\bar{u}$. Due to the special form (2.5) of $\mathcal{R}$, we note in this case that

$$(2.6) \qquad\qquad \mathcal{R}(0) = 0 \quad \text{and} \quad D\mathcal{R}(0) = 0.$$

These properties in turn imply that the inertial manifold contains 0 and it is tangent at 0 to $\mathcal{P}_m H$. Note that 0, in this case, is a stationary solution of (2.2) and, like any stationary solution, it belongs to the attractor included in the inertial manifold.

Equation (2.2) in the context (2.4)-(2.6) occurs very precisely in meteorology when (2.1) are the meteorology equations (of Navier-Stokes type). In this context Theorem 2.1 below produces the first rigorous proof concerning the existence and the properties of slow manifolds.

Beside the hypotheses above and some technical hypotheses concerning $A, R, \mathcal{A}$ and $\mathcal{R}$ we assume that

$$(2.7) \qquad\qquad \lambda_j \sim c\, j^p \quad \text{as } j \to \infty,$$

for some p satisfying

$$(2.8) \qquad\qquad p(1 - \alpha) > 1$$

and we make the following hypothesis directly related to the spectral gap hypothesis:

$$(2.8) \qquad\qquad
\begin{aligned}
&\textit{For every constant } c \textit{ there exists an} \\
&\textit{arbitrary large } m \textit{ such that} \\
&\lambda_{m+1} - \lambda_m > c\left(\lambda_{m+1}^\alpha + \lambda_m^\alpha\right).
\end{aligned}$$

We have

THEOREM 2.1. *Under the above hypotheses, for a sufficiently large m, there exists a $\mathcal{C}^1$ mapping from $\mathcal{P}_m H$ into $\mathcal{Q}_m H$ whose graph $\mathcal{M}$ is an inertial manifold for equation (2.2).*

If (2.6) is satisfied, then $\mathcal{M}$ contains 0 and is tangent at 0 to $\mathcal{P}_m H$. In particular in the context (2.4)-(2.6) $\mathcal{M}$ is an inertial manifold for equation (2.1) as well.

This result is proved in Debussche-Temam (1991). A particularly delicate part of the proof is the establishment of boundedness properties of the spectral projectors $\mathcal{P}_m, \mathcal{Q}_m$, uniform with respect to m.

For the definition and the utilization of inertial manifolds in meteorology see e.g. Daley (1981), Tribbia (1979, 1982) (and Temam (1990b)).

3. Incremental unknowns. As indicated in Section 2, inertial manifolds and approximate inertial manifolds are usually derived as graphs of a function Φ for $P_m H$ into $Q_m H$; $P_m H$ is the space spanned by the first m eigenvectors of A, $w_1, ..., w_m$ (see (2.3)), P_m the orthogonal projector onto $P_m H$ and $Q_m = I - P_m$. If u belongs to H, then we write

(3.1)
$$\begin{cases} u = y_m + z_m, \\ y_m = P_m u, \quad z_m = Q_m u, \end{cases}$$

and

(3.2)
$$z_m = \Phi(y_m),$$

is the equation of the (approximate/exact) inertial manifold. Since $P_m H$ is associated to the small wave-numbers ($\lambda_1 \le \lambda \le \lambda_m$) and $Q_m H$ is associated to the large wave-numbers ($\lambda \ge \lambda_{m+1}$), y_m and z_m can be naturally considered as the large waves and small waves components of u. Hence equation (3.2) represents the slaving of the small modes by the large ones.

For infinite dimensional dynamical systems, in particular those associated to partial differential equations, the usual method of approximation include spectral methods (and among them those using the empirical eigenfunctions, see e.g. Sirovich (1990)), finite differences and more recently wavelets. Decompositions into small and large waves like (3.1) occur naturally when spectral methods or wavelets are considered. However when a partial differential equation is discretized by finite differences there is no natural decomposition between small and large waves : indeed in this case the unknowns are the nodal values of the function and essentially all nodal values play the same role.

In order to overcome this difficulty and to be able to build approximation of inertial manifolds in the context of finite differences, we have introduced in (Temam (1990a)) the concept of Incremental Unknowns.

For the definition of incremental unknowns at least two nested grids are necessary like in multigrid methods. We call them the coarse grid and the fine grid. The incremental unknowns consist then of the following :

- The nodal values of the function at the coarse grid points and

- At the fine grid points not belonging to the coarse grid points the incremental unknown is the increment to the averaged value of u at the closest coarse grid points.

Of course by Taylor's formula the incremental values at the fine grid points are expected to be small. We denote by Y the properly ordered set of nodal values at the coarse grid points and by Z the properly ordered set of incremental values at the fine grid points.

The procedure can be reiterated when more than two nested grids are used. In that case Y denote the set of nodal values of the function at the coarsest grid and Z the set of increments at the successive grids. Assume that h_0 is the mesh for the coarsest grid and $h_j = h_0 2^{-j}$ the mesh on the level j; just by Taylor's formula the components of Z at level j decay exponentially in magnitude, like $(h_0/2^j)^2$.

Consider first a self-adjoint linear elliptic problem. After discretization by finite differences with mesh h we obtain a linear system

$$(3.3) \qquad\qquad AU = b,$$

where $U, b \in \mathbf{R}^N$, A is a symmetric positive definite matrix. The vector U corresponds to the nodal values of the unknown function u, say $u(\alpha h, \beta h)$ in $\mathbf{R}^2$. The introduction of the incremental unknowns corresponds to a **very simple** change of variable

$$(3.4) \qquad\qquad U = S\bar{U}$$

where S is the transfer matrix and $\bar{U} = \begin{pmatrix} Y \\ Z \end{pmatrix}$ can be split into the coarse grid component Y and the fine grid(s) component Z (namely the increments).

With (3.4), (3.3) becomes

$$AS\bar{U} = b$$

and after multiplication by ${}^t S$:

$$(3.5) \qquad \begin{aligned} \bar{A}\bar{U} &= \bar{b} \\ \bar{A} &= {}^t SAS, \quad \bar{b} = {}^t Sb. \end{aligned}$$

The numerical solution of systems like (3.5) is studied in Chen-Temam (1991a,b,c) (see also Garcia (1991)). It is shown in the first reference that, for linear elliptic problems, the IU method is as efficient as the classical V-cycle multigrid method.

In the case of dissipative evolution equations, the study in Temam (1990a) leads to approximate inertial manifolds of equation

$$Z = \Phi(Y).$$

From the strict numerical viewpoint the advantages of the IU method are the following :

- Since the Z components of the incremental unknowns are small there is a large set of small unknowns whose determination need not be done with too much accuracy and this leads to important improvements in computational efficiency.

- When the equations are discretized in space and time the utilization of the incremental unknowns improves the stability condition for the time step ; this allows the utilization of a larger time step and leads also to important savings in computing time.

The stability analysis of the nonlinear Galerkin method and of the IU method are studied in Temam (1991).

It is noteworthy that spectral methods produce a decomposition u of the form

$$(3.6) \qquad\qquad u = y + z,$$

where y and z are small and large wave in the Fourier (wave- number) space. With finite differences and incremental unknowns y and z are large and small waves in the physical space. Of course we expect wavelets to produce a decomposition of u into small and large waves both in the physical and in the Fourier space.

Other forms of incremental unknowns are studied at this time, in particular oscillatory type incremental unknowns similar to wavelets. This work will appear elsewhere.

REFERENCES

A.V. BABIN AND M.I. VISHIK, *Attractors of partial differential evolution equations and their dimension*, Russian Math. Survey, 38 (1983), pp. 151–213.

M. CHEN AND R. TEMAM, *The Incremental Unknown Method I, II*, Applied Math. Letters (1991a).

M. CHEN AND R. TEMAM, *Incremental unknowns for solving partial differential equations*, Numerische Math. (1991b) (to appear).

M. CHEN AND R. TEMAM, *Incremental unknowns in finite differences : condition number of the matrix* (1991c) (to appear).

S.N. CHOW AND K. LU, *Invariant manifolds for flows in Banach spaces*, J. Diff. Equations, 74 (1988), pp. 285–317.

P. CONSTANTIN, C. FOIAS, O. MANLEY AND R. TEMAM, *Determining modes and fractal dimension of turbulent flows*, J. Fluid Mech., 150 (1985), pp. 427–440.

P. CONSTANTIN, C. FOIAS, B. NICOLAENKO AND R. TEMAM, *Integral Manifolds and Inertial Manifolds for Dissipative Partial Differential Equations*, Springer-Verlag, New York, Applied Mathematical Sciences Series, vol. 70., 1988.

P. CONSTANTIN, C. FOIAS, B. NICOLAENKO AND R. TEMAM, *Spectral Barriers and Inertial Manifolds for dissipative partial differential equations*, J. Dynamics and Differential Equ. 1 (1989), pp. 45–73.

P. CONSTANTIN, C. FOIAS AND R. TEMAM, *On the dimension of the attractors in two-dimensional turbulence*, Physica D 30 (1988), pp. 284-296.

R. DALEY, *Normal mode initialization*, Rev. of Geoph. and Space Physics 19 (1981), pp. 450-468.

C. DOERING AND J.D. GIBBON, *A note on the Constantin-Foias-Temam attractor dimension estimate for two-dimensional turbulence*, Physica D 48 (1991), pp. 471–480.

C. FOIAS, G.R. SELL AND R. TEMAM, *Variétés inertielles des équations différentielles dissipatives*, C.R. Acad. Sci. Paris, Série I, 301 (1985), pp. 139–142.

C. FOIAS, G.R. SELL AND R. TEMAM, *Inertial manifolds for nonlinear evolutionary equations*, J. Diff. Equ. 73 (1988), pp. 309–353.

C. FOIAS AND R. TEMAM, *Some analytic and geometric properties of the evolution Navier-Stokes equations*, J. Math. Pures Appl. 58 (1979), pp. 339–368.

S. GARCIA, *article in preparation* (1991).

J.M. GHIDAGLIA AND R. TEMAM, *Lower bound on the dimension of the attractor for the Navier-Stokes equations in space dimension 3*, in Mechanics, Analysis and Geometry : 200 years after Lagrange, M. Francaviglia Ed., Elsevier, Amsterdam, 1991.

I.C. GOHBERG AND M.G. KREIN, *Introduction to the Theory of Linear non Selfadjoint Operators*, Translations of Mathematical Monographs, vol. 18, A.M.S. (1969).

Y. POMEAU AND P. MANNEVILLE, *Stability and fluctuations of a spatially periodic convective flow*, J. Phys. Lett. 40 (1979), pp. 609–612.

L. SIROVICH, *Empirical eigenfunctions and low dimensional systems, Center for Fluid Mechanics, Brown University*, Preprint, n° 90–102 (1990).

L. SIROVICH, B.W. KNIGHT AND J.D. RODRIGUEZ, *Optimal low dimensional dynamical approximations*, Quarterly of Appl. Math., 48 (1990), pp. 535–548.

R. TEMAM, *Navier-Stokes Equations*, North-Holland Pub. Company, 1977.

R. TEMAM, *Attractors for Navier-Stokes equations*, in *Nonlinear Partial Differential Equations and their Applications*, Séminaire du Collège de France, vol. VII, H. Brezis, J.L. Lions Eds., Pitman, 1985.

R. TEMAM, *Inertial manifolds and multigrid methods*, SIAM J. Math. Anal. 21 (1990a), pp. 154–178.

R. TEMAM, *Inertial Manifolds*, The Mathematical Intelligencer 12 n° 4 (1990b), pp. 68–74.

R. TEMAM, *Stability analysis of the nonlinear Galerkin method*, Math. of Computations (1991) (to appear).

J.J. TRIBBIA, *Nonlinear initialization on an equatorial Beta-plane*, Mon. Wea. Rev. 107 (1979), pp. 704–713.

J.J. TRIBBIA, *On variational normal mode initialization*, Mon. Wea. Rev. 110 (1982), pp. 455-470.

J.J. TRIBBIA, *A simple scheme for high-order normal mode initialization*, Mon. Wea. Rev. 112 (1984), pp. 278-284.